ELECTRONIC COMMUNICATION

SAMI ZLITNI & FABIEN LIENARD (ed.)

ELECTRONIC COMMUNICATION

POLITICAL, SOCIAL AND EDUCATIONAL USES

PETER LANG
Bern · Berlin · Bruxelles · Frankfurt am Main · New York · Oxford · Wien

Bibliographic information published by die Deutsche Nationalbibliothek
Die Deutsche Nationalbibliothek lists this publication in the Deutsche Nationalbibliografie; detailed bibliographic data is available on the Internet at ‹http://dnb.d-nb.de›.

British Library Cataloguing-in-Publication Data: A catalogue record for this book is available from The British Library, Great Britain

Library of Congress Control Number: 2015942735

With the cooperation of the IUT of Le Havre.

ISBN 978-3-0343-1687-3 pb. ISBN 978-3-0352-0321-9 eBook

This publication has been peer reviewed.

Contents

ICT and political communication

Education, Identity and Electronic Communication

Fabien Liénard, Sami Zlitni & Marina Haan

Electronic Communication in Digital Societies

Following the line of the previous collective publications (Liénard and Zlitni, on 2011, 2012; Zlitni and Liénard, on 2012, 2013), this volume deals with what we consider to be an object of multidisciplinary research: electronic communication. Patrick Charaudeau finely illustrates the idea of multidisciplinarity, meanwhile, his views engender a brief reflection that we intend to bring forward:

> I would like to avoid the fault of pretending to be a specialist of all these disciplines [...], and even though, for the sake of the idea of interdisciplinarity which I advocate, I read, as much as possible, the works of historians, philosophers, sociologists, anthropologists, psychologists and the specialists of other disciplines, I cannot claim that the degree of my knowledge is sufficient for me to discuss everything (2012 : 172).

We thus read, as much as possible, the works of the researchers of different disciplines dealing with the same object of scientific interest: electronic communication. But we agree with Patrick Charaudeau that this experience cannot turn us into specialists of these disciplines. It is thus essential to apply to the researchers specializing in sciences of information and communication, as well as in language sciences (linguists, sociolinguists, psycholinguists or didacticiens, etc.), in sciences of education, political sciences, behavioral sciences, social sciences, sciences of management, etc. We mean that the complexity of the scientific object enhances the importance of interdisciplinarity and accounts for the multidisciplinary logic of this work.

Numerous authors discussed interdisciplinarity, multidisciplinarity, and even transdisciplinarity, polydisciplinarity. Among them there is Claudine Blanchard-Laville (2000), Yves Le Noir and Lucie Sauvé (1998), Bernard Valade (1999), Dominique Vink (2000) or, evidently, Edgar Morin (1994, 1999). First of all, we shall refer to the ideas of the latter, namely, to his reminding us of the fact that a discipline is:

> An organizational category within the scientific knowledge [which] establishes the division and the specialization of work and corresponds to the diversity of the domains covering the sciences [...]. A discipline has a natural tendency towards autonomy by means of demarcating its boundaries, establishing its vocabulary, creating or applying its techniques and, possibly, by developing its own theories (1994)[1].

But in the same article, he specifies the fact that the history of disciplines is made out of

> [...] breaks in the disciplinary boundaries, the encroachments of a problem of one discipline upon another, the circulations of concepts, the emergence of hybrid disciplines which will end up becoming autonomous.

Without pushing the reasoning up to this paroxysm, we can admit that the peculiarities of even electronic communication make the sciences (human, social, as well as technical) collaborate in order to observe (and better identify) the plurality of the practices and the uses of electronic communication. This fact makes us consider the multidisciplinary and interdisciplinary cases to be likely to articulate the cooperations when the research object is common. Indeed, considering what Alain Glykos says (1999) allows us to admit that multidisciplinarity is "An association of disciplines which contributes to a common achievement, but without which every discipline has to modify its own vision of things and its own methods" (in Claverie, on 2009: 14). The limits of this approach are obvious and particularly explicit in the situations with a low degree of sharing information because the researchers of every discipline "return" to the principles of their respective sciences as soon as they have finished the study requiring a disciplinary link: the multidisciplinarity. But, finally, it is does not refer to interdisciplinarity either, because, as Pamela Mitchell explains, this phenomenon results in regrouping "the researchers working together, each of them working on their disciplinary groundwork in order to solve the common problem" (2005: 333). The great difference, according to Edouard Kleinpeter (2011), lies in the fact that in the latter case "the scientists would not work anymore in parallel [like in multidisciplinary situations], but would rather communicate with each other on a regular basis". We, undoubtedly, put this suggestion in the highlight of the actual work (it also having been the guiding line of our scientific research since 2011): we offer texts refering to particular sciences

1 This article by Edgar Morin, consulted in Octobre 2014, is available at the following address : <http://perso.club-internet.fr/nicol/ciret/bulletin/b2c2.htm>.

in a multidisciplinary logic, in the same way as the chapters of this book contribute to the scientific decompartmentalization from an interdisciplinary standpoint. In other words, we participate in the transverse exchanges between researchers, who, by means of their specific concepts and methods, suggest observing, studying the same uses and the same practices of electronic communication.

As a matter of fact, every day electronic communication becomes a little bigger integral part of everyday life of billions of human beings. One of the explanations lies in the technical and technological quality of equipment (structural and personal), regardless of our geographical location. The mediating tools are based on more and more technologies while getting much simpler to use. They also turn out to be highly complementary. The complementarity and the convergence prove to be two assets, two forces, which undeniably participate in their development and allow them to attract a large number of users. Speaking about complementarity implies admitting the fact that, even if the competition between various economic agents of ICT (manufacturers, developers, distributors, etc.) is fierce, the acquisition of a computer and an intelligent telephone (or a smartphone), for example, makes it possible to cross the uses without facing real problems of compatibility or configuration. This technological complementarity thus goes hand in hand with an increasing convergence: the uses become more diversified (the users can perform more and more operations with the same ICT), there are new practices being developed and converging toward the communication modes that just not long ago required the employment of a specific ICT. Nowadays it is possible and very simple to browse the Internet, to consult one's e-mails or a Facebook profile, to send SMS or to publish a video from a computer or from a mobile phone. The programmes and the communication interfaces are logically accomplished to stimulate this convergence and to encourage the users to employ ICT constantly. Thus, the notion of *connected presence*, highlighted by Christian Licoppe reveals all its dimension, all its sense:

> the relationship between friends and relatives becomes a dense, seamless tissue, made by means of continuous interlacing of face-to-face interactions and of popularizing through the media acts of communication, the proliferation of the contacts being a guarantor for the link. This connected presence blurs the borderline between presence and absence, because due to linking up the participants are always a little bit present in others (on 2009: 30).

The relationship between the individuals is continuous, it being prolonged from face-to-face communication to the exchanges via ICT. The frequency of contacts and digital exchanges increase meanwhile the time of interaction, of speaking, decreases. Christian Licoppe gives an example of phone calls which become more frequent and shorter means of communication, but we can also point out the success of SMS to illustrate the idea. The explanation is simple: the use of ICT takes root in our everyday life and integrates into daily human activities for (and to the point of) perpetuating the face-to-face relationship without time limit. There is no more pressure, no necessity of exchange and finally ICT prolong physical relationship in time, being no more attribute of something exceptional or important but an element of everyday routine and mundanity. Christian Licoppe specifies the idea by indicating that

> the interpersonal technologies of communication (a landline phone, a mobile phone, voice and electronic mails, SMS, handwritten correspondence, etc.) build up the resources for establishing contacts, in the same way as face-to-face means of communication do. They are like numerous threads which, all together, weave the relational tapestry (ibidem: 32).

This relational tapestry, made out of a tangle of electronic messages, provides structurizing more or less stable discourse spheres (i.e., regularly frequented by interagents) according to the convenience of the (virtual) place, the theme, the exchanges or the leaders. It means that we consider the alleged discourse spheres to be virtual communities, that are a set of social entities connected with each other through social and verbal digital interactions carried out and maintained with a precise purpose, being shared by all the users, all the members of the network concerned. As soon as the theme, the purpose is identified, an exchange "programme" gets more or less explicitly developed and interaction rites are worked out. They vary a lot in nature and in the quality of the exchanges in focus, which is likely to depend on the parameters mentioned above: the quality of the place, the exchanges and the leaders. To specify this idea, we resort to the works of Anita Blanchard who actually explains: "Virtual behavior settings are created through the shared interactions of members and a developing sense of space or 'place' in CMC" (2004: 2). Inspired by the works of Jones (1997)[2], she compares the functioning of a social network (thus representing a digital discourse sphere) with that of a café. Indeed, the social structure of such an establishment is pyramidal with the chief manager at its summit, that is the café owner who is the coordinator of the place and, above all, a decision-maker, the fact accounting for his hyper-activity. He requires actions from his staff (waiters and other employees) who can also have coordination responsibilities to satisfy the needs of the individuals constituting the third level: the customers. Anita Blanchard distinguishes between several types of customers. The "true" customers are the soul of the place: they make it live and identify themselves with it to various extents, all of them expecting a lot of hierarchical levels in return, according to their loyalty (their attachment to the establishment). Occasional customers and passers-by (the ultimate level) do not have the same expectations. The representatives of the first category look for the satisfaction of their primary needs (finding a shelter, having something to eat or getting relief). The others are eager to discover an environment, an ordinary discourse space. All these members interact socially and verbally more or less intensively, the most involved ones (at Levels 1, 2 and 3) participating undoubtedly more (in terms of frequency, but also in terms of intensity). So Anita Blanchard attracts the readers' attention to the virtual communities which she researches more specifically. She explains that: "Participants in virtual behavior settings can be characterized by activity level, ranging from leaders to lurkers" (2004: 13). To specify her point of view Anita Blanchard has created the table reproduced by us bellow:

2 Jones develops a "cyber-archaeological" approach and explains that an archaeologist observing the vestiges of a village will identify the obvious traces of the Man's existence, the human activity which will be informative in terms of the interrelations carried out and, thus, indicating the possible interactions between the members of this group, of this community. Applied to digital social networks, the approach seems to be interesting.

Category	Participant type	Definition
Leader	Technical leader	Sees that program that supports CMC forms runs. Also known as moderator (asynchronous) and wizard (synchronous).
	Information leader	Major provider of expertise and knowledge about a topic.
	Social leader	Major provider of social support among members.
Participant	Participant	Contributes along a continuous range of communication.
Lurker	Private communicator	Communicates to other participants through private email.
	Never communicates	Never communicates to other participants or to community.

The table illustrates how much the *tapestry*, referring to Christian Licoppe's metaphor, is elaborated and even complicated. It can be accounted for the fact that the spheres, the virtual communities overlap, intercross and become entangled thanks to the users who change their communication environment and whose electronic discourse participates in the construction of digital identities at the same time.

Fanny Georges is one of the French specialists of digital identities. According to her, in a situation of electronic communication (no matter if the user of ICT is sending an e-mail or is having a discussion in a social network) there are three identity levels which we are going to present in brief. At first, the user has a *declarative identity* which he voluntarily demonstrates in the digital environments: it is what he gives for the others to see. Then the user takes an *active identity* when he is going to take the tool and to develop communication practices in order to produce some particular effects: it is what he gives to see willingly or unwillingly while performing some concrete actions which contribute to delimiting the outlines of what he represents on the Web. Finally, the user establishes a *calculated identity* when he tends to develop the actions or a discourse that will enable him to create a particular image: it is what the user wishes the others to see of him… These three identities work successively, as well as simultaneously, to finally establish the user's personal representation and *digital identity* (Georges, 2009). Dominique Cardon (2008), Julie Denouël (2011) and Louise Merzeau

(2009) are among the other researchers dealing with the digital identity. Besides, the latter prefers speaking about the *digital presence*, thus, approaching the position of Christian Licoppe. In fact, Louise Merzeau explains that nowadays the digital identity can be characterized through the analysis of the digital traces that the ICT user is sure to leave alongside, throughout all his electronic practices. Above all, it can be characterized thanks to the accomplished uses, the frequency, the digital presence. So,

> "The digital identity thus acquires, by itself, a market value: it can be bought and sold in the form of behavioural advertisements and business files" (on 2009: 3). She also adds: "For the user the attraction of blogs, sharing information platforms and social networks consists of putting his or her individual characteristics on the common ground (centers of interest, preferences, comments, statuses, etc.)" (ibidem).

And if the user recognizes himself so much in the discourse sphere frequented by him assiduously, it is because this sphere also integrates the traces incorporated there beforehand (thus at the level of the active identity revealed by Fanny Georges). Actually,

> The user indeed learns to interpret the evaluation of his identity in terms of influence or reputation and adjusts his signals to stick to the compatibility model valued by the social Web (ibidem).

Such statements encourage Louise Merzeau to put forward the idea of a digital double because finally the digital identity, if it can be modified or even fantasized about, is in most cases the true expression of what the user is fundamentally. The explanation is simple: "the digital environment is a memory 'by default' " (ibidem: 10). The following chapters integrate this dimension and the parameters that we have just touched on. They result from the works of the researchers specializing in the fields of electronic communication and studying, from different points of view, the questions of ICT, the ensuing political, social and educational uses and the emerging practices. But we believe that the presentation of the crossed approaches and the points of view, which are sometimes divergent, ensures, first of all, a better awareness and understanding of electronic communication, its significance, stakes and the limits for the digital societies worldwide, on almost all the continents.

The work opens with Yves Winkin's prolegomena entitled "Do we really need to fly from communicational utopia?" It is a text of a plenary conference presented by the professor at the international conference

"Electronic communication, culture and identities" which took place at the Information-Communication Department of the IUT of Le Havre in June, 2014. Contrary to Breton Philippe, insisting in his key work on the necessity of going out of the utopia of communication, Yves Winkin wonders in his text if the utopia of communication, understood as pooling, does not deserve to be protected. The latter brings forward the argument that injecting the utopia of giving / receiving/ returning in the contemporary social life is probably not such a bad utopia.

So these prolegomena creates a link between the anthropology of communication and the uses of electronic communication. And it is for highlighting the richness of this question that we have constructed this work of two multidisciplinary parts. The first part is devoted to the uses and the practices of ICT in the political domain. The reader will find the chapters dealing with the use of social networks at elections and those revealing the use of ICT as tools of mobilization and social structuring. Besides the political domain, the electronic communication can play an important role in some learning processes developed by or for the learners. Thus, the use of ICT in the field of education and in the construction of identities is tackled in the second part of this work.

ICT and political communication

The increase of ICT stimulated the development of electronic communication in the political domain. It does not matter who the actor of communication is: a citizen or a candidate for elections, various electronic means of communication allow each of them to have access to new platforms of expression and to be more visible in the public space, providing everyone with independence by comparison with the traditional media. The first chapter is devoted to the use of social networks during the Indian general elections of 2014. In fact, Pitabas Pradhan ("Social Media Impact on Election Campaign: A Study of the Indian General Elections 2014") states that the growing influence of social media has forced the mainstream political parties to reframe their media plans. Social media tools are used along with the old media forms, which used to play a key role in political communication for long. The social media based political communication efforts have not only changed the modes of message delivery but also perceptibly changed the nature of political discourse in the country. The social media platforms have given the citizens access to new public forums to convey ideas and carry political actions to the convenience of the masses at a far faster speed. These tools present better opportunities for politicians to have more visibility in the public sphere and propagate alternative political agenda. The new ICT tools provide unrestricted opportunities to the citizens to express themselves, unlike the mainstream media where the gate keeping theory still exists. The impact is so huge that the social media political discourses are shaping the mainstream media stories to a greater extent. Against this backdrop, Pitabas Pradhan attempts to explore the strategic use of social media by mainstream political parties during the Delhi Assembly elections 2013 and the general elections 2014. This study was conducted in the backdrop of the growing perception that a one-year old party claimed a sensational victory over the mainstream parties on the strength of its social media strategy. The author builds this study on media content analysis, social survey of the citizens, and in-depth interview of experts to explore the role and impacts of social media in election campaigns and its outcome.

Chapter 2 by Zhe Deng and Richard Balme ("Chinese Micro-Bloggers' Behaviors towards Twiplomacy. Text Analysis of British PM David Cameron's Micro-blogs") is devoted to exploring the communication and interaction mechanism of Twitplomacy on micro-blog, the Chinese version of Twitter. By text analysis of 200 pieces of comments on each of Cameron's twelve micro-blogs, and investigation on the micro-bloggers' characteristics, the objective of the authors is to categorize Chinese netizens' behaviors of using micro-blog, so as to show what kinds of information that the citizens actually want to know, to express or even to diffuse. Also they analyze the correlation between micro-bloggers' backgrounds or characteristics and their online behaviors, thus the result could to some extent be guidance for the governments and political leaders of their ICTs usage and regulation in view of different network environment and communication situation. For Zhe Deng and Richard Balme, China's micro-blog provides a basic public energy field for Chinese engagement, referring to its information service to common people and allowance of all micro-bloggers' participation. However, the difference between China's micro-blog and real public energy field is still obvious from the attitudes and behaviors of Chinese micro-bloggers.

As for Aysun Akan ("Making Sense of Gezi Park Protests: Gezi Park Protests as a Defence of Lifestyles"), she expresses her interest in the use of social networks as tool of protest and social mobilization by studying the case of the the Gezi Park protest movement. Indeed, protests in Istanbul, and fifty other cities in Turkey, in May and June 2013 reflected a deep ideological polarisation in Turkish society, between conservatives (read Islamic) and modern (read Western), liberal Turks. One of the dominant themes in the protests was the criticism targeting, then the Prime Minister, Tayyip Erdoğan's, the leader of the JDP's, Justice and Development Party, (Adalet ve Kalkınma Partisi, AKP), interference in life styles different than an Islamic life style. The fear of Islamisation of society seats deep in the minds of secular Turks. From the beginning of the modern Turkish Republic (1923), throughout the first half of the twenty century, political, military and state elite prevented political or civil activities with Islamic leaning. Popularity and the electorate success of the JDP and the increasing authoritarian tendency of Mr. Erdoğan and his intolerance towards diverse life styles triggered the fear of Islamisation of society. The Gezi Park protests, among other things, reflected this battle of identities in the widely circulated slogans, messages and images through the social media.

Within the frames of the epistemological field of research, Abhay vir Singh Kanwar's article ("Theorizing Cyber activism: Drawing from Durkheimiam lenses") deals with the problem of cyber activism. The paper attempts to delve into intricacies of cyber space and expostulate how it is not only mapping, gauging and molding the various strata's of society but also cutting across different type of diversity of people across globe. In this chapter, Abhay vir Singh Kanwar explores how cyberspace reproduces the attributes of "social fact", one of the eminent conceptions of Durkheim. Collective conscience which ensues into social integration which establishes primacy of society over individuals is also one of his prepositions. Also, the study focuses on the impact of cyber grid on society's collective conscience.

Education, identity and electronic communication

The development of ICT and their integration in the educational methods have changed the way the learners get knowledge and have modified considerably the organization of teaching. So the primary objective of the second part of the work is to highlight the uses and practices of ICT by learners and teachers. It also reveals the way the ICT contribute, in a voluntary or unconscious way, to the construction of new identities.

The first chapter of the second part deals with a case of American students. Indeed, Alain-Philippe Durand and Elyse Petit ("The Teaching of French and Digital Humanities") analyze the electronic uses and the interactive collaboration between the students and Jean-Philippe Toussaint, known as a novelist, a film-maker and a Belgian photographer. Indeed, between 2000 and 2010 at the University of Rhode Island and since 2010 at the University of Arizona, Toussaint participated in several activities in Durand's seminars on the contemporary novel and film: dialogues with students on the Internet, blogs, on Skype, or face to face during transatlantic visits. Encouraged by these experiences, Toussaint invited Durand to partner with him on what could be a unique project in contemporary literature: to coordinate and maintain in an interactive way, along with his students, the American page of the Jean-Philippe Toussaint web site (jptoussaint.com). This chapter recounts and presents examples of this collaboration between Toussaint, Durand, and his students.

Another case study, Greek and Finnish this time, enabled Marianna Vivitsou et Kirsi Viitanen ("The pedagogies of the future: through young people's eyes in storytelling experiences with the digital in Finland and Greece") to tackle the issue of changes in teaching methods as a result of the use of electronic communication. In this way, by allowing a different 'language' to come up through image-based practices for movement and change, social media seem to create a new pedagogical space. When social media are used for pedagogy and learning with image-based practices, it is in that new space where pedagogy can transform. Marianna Vivitsou et Kirsi Viitanen hold the position that pedagogies with the digital, or without it, cannot be transformative unless the focus shifts so that, in addition to the (digital) pedagogue, it includes the youth as decision makers and as curriculum planners. In the pedagogies of the future the young people are leaders, partners, and architects of their own learning. In this paper the authors argue, by discussing storytelling experiences with the digital of young adolescents from Finland and Greece, that such shift is feasible. To support their case, they examine how these young people developed initiative and action as storytellers and filmmakers in collaborative physical and virtual spaces.

Besides, these uses of ICT, getting more and more numerous, lead to particular electronic practices and can transform the scriptural practices of young people. It is in this sense that Olga Volckaert-Legrier, Josie Bernicot, Antonine Goumi, and Alain Bert-Erboul ("Text Messages: Enemy or Ally in the Spelling Learning Process? A Longitudinal Study of 11–12-Year-Old Junior High School Students") dealt in their chapter with the question of "textism" in schoolchildren's SMS messages. The authors analyzed a corpus of 4,524 text messages produced in daily life situations by 19 junior high school students, all of whom were 12 years of age. The participants had never owned or used a mobile phone before beginning the study, and their text messages were collected each month over the course of one year. To characterize the text messages, the density of textisms was used, defined as the number of changes in a word's spelling form as compared to traditional spelling divided by the total number of words in the text message. One addition was made to the existing research: the distinction between textisms that were consistent with the traditional code and textisms that broke with that code – in other words, those that were the most typical of text-message language. The analysis of the results showed that the only existing link between the level of spelling and the density of textisms was with regard to textisms breaking with traditional code during the first trimester of text-message use. Contrary to widespread fears, students with a strong level of spelling were the ones who made numerous textisms breaking with traditional code, and conversely. This study showed that traditional writing and text-message writing are dependent upon the same symbolic cognitive abilities. Furthermore, this study arrived at the conclusion that text messages are not a threat to the level of spelling at school, but instead constitute a new and additional opportunity for writing practice.

Sara Minucci and Marta Severo ("Strategies for building online identities in academia. An exploration of digital communication of researchers in social sciences and humanities") are interested in the digital communication of researchers in social sciences. They focused especially on the three facets of the use of the web in academia: how researchers use the web and social platforms to promote their personal and professional image; how researchers use the web and social platforms to build their online identities; and which relationship does exist between the image that the researcher intends to build and the image that factually appears on the web. The empirical analysis has been developed in three phases. Firstly, an online questionnaire has been administered to OMERA subscribers to understand which is the relation between researchers and digital tools both for professional and personal communication, and which strategies researchers adopt for building their online identity. Secondly, the factual online image of the respondents has been analysed through an ethnography of the internet. Thirdly, the hypothesis that the different behaviours of researchers for building their online identity can be reduced to three main strategies has been tested through in-depth interviews to a sample of users that answered to the OMERA questionnaire.

In light of the unresolved debates surrounding data usage and protection, Danielle Masterton ("Identity in the Era of Big Data: An Embodied Approach") proposes an embodied approach as an alternative framework for conceptualising identity, and its value, in the context of big data. Merleau-Ponty's phenomenology will be used to inform a model of embodiment that can be leveraged against disembodied visions of data and capitalist ideology perpetuated by industry and standard-setting institutions. The World Economic Forum (WEF) conceptualises data as an asset, relying on the alienation of data from its producer in order to attain a conceptualisation whereby data circulates freely as a commodity. The World Economic Forum's proposal will be juxtaposed with an embodied approach, inspired by the European Commission's perspective in which personal data is considered as part of the self. Staging an encounter between these opposing perspectives will bring to light the contradictions at the core of assumptions about big data. The complexity of the relationship between data, the body, identity and commodity capitalism is not addressed by industry discourses and therefore merits further investigation.

We thank all the researchers who contributed to this work and thus participated in a better understanding of political, social and educational uses of electronic communication, no matter that some authors spoke about numerical communication or about digital communication or about virtual communication.

References

Blanchard, Anita, 2004, "Virtual Behavior Settings: An Application of Behavior Setting Theories to Virtual Communities", in *Journal of Computer-Mediated Communication*, 9(2). Retrieved from <http://jcmc.indiana.edu/vol9/issue2/blanchard.html>.

Blanchard-Laville, Claudine, 2000, "De la co-disciplinarité en sciences de l'éducation", in *Revue française de pédagogie*, n° 132.

Cardon, Dominique, 2008, "Le design de la visibilité : essai de cartographie du web 2.0", in *Réseaux*, n° 152.

Charaudeau, Patrick, 2012, "Pour une interdisciplinarité "focalisée". Réponses aux réactions", in *Questions de Communication*, n° 21, Nancy, PUN.

Claverie, Bernard, 2009, "La transdiciplinarité : à travers les réseaux du savoir". Retrieved from <http://documents.irevues.inist.fr/bitstream/handle/2042/28893/SurlatransdisciplinariéA9.pdf>.

Denouël, Julie, 2011, "Identité", in *Communication*, n° 88, Paris, Seuil.

Georges, Fanny, 2011, "L'identité numérique sous emprise culturelle. De l'expression de soi à sa standardisation", in *Les cahiers du numérique*, n° 7 (1).

Glykos, Allain, 1999, *Approche communicationnelle du dialogue artiste/scientifique*, Note de synthèse pour la HDR. Université Paris 7.

Kleinpeter, Edouard, "Pour une définition de l'interdisciplinarité", communication tenue à l'occasion d'une Journée de l'interdisciplinarité organisée par l'Institut des Sciences de la Communication du CNRS (ISCC), 29 juin 2011, Paris.

Le Noir, Yves et Sauvé, Lucie, 1998, "De l'interdisciplinarité scolaire à l'interdisciplinarité dans la formation à l'enseignement : un état de la question", in *Revue Française de pédagogie*, n° 124.

Licoppe, Christian, 2009, *L'évolution des cultures numériques*, Paris, Edition Fyp.

Liénard, Fabien & Zlitni, Sami, 2011 (dir.), *La communication électronique : enjeux de langues*, Limoges, Editions Lambert-Lucas.

Liénard, Fabien & Zlitni, Sami, 2013 (coord.), *Traces numériques : de la présence à l'oubli*, in *NETCOM* (Networks and Communication Studies), n° 26, 1/2, Retrieved from <http://www.netcom-journal.com/>.

Merzeau, Louise, 2009, "Présence numérique : les médiations de l'identité", in *Les Enjeux de l'information et de la communication*. Retrieved from <http://w3.u-grenoble3.fr/les_enjeux>.

Mitchell, Pamela H., 2005, "What's in a name? Multidisciplinarity, Interdisciplinarity, and Transdisciplinarity", in *Journal of Professional Nursing*, n° 21 (6).

Moatti, Alexandre, 2012, "Le numérique : adjectif substantivé", in *Débat*, n° 170, Paris, Gallimard.

Morin, Edgar, 1999, "Relier les connaissances : le défi du XXIe siècle", Paris, Editions du Seuil.

Morin, Edgar, 1994, "Sur l'interdisciplinarité", in *Bulletin interactif du Centre international de recherches et études transdisciplinaires*, n° 2.

Valade, Bernard, 1999, "Le 'sujet' de l'interdisciplinarité", in *Sociologie et sociétés*, vol. XXXI, n° 1. Retrieved from <http://www.erudit.org/erudit/socsoc/v31n01/valade/valade.htm>.

Vinck, Dominique, 2000, *Pratiques de l'interdisciplinarité. Mutations des sciences, de l'industrie et de l'enseignement*, Grenoble, PUG.

Vygotski, Lev S., 1985, *Pensée et langage* (3ème édition en 1997), Paris, La Dispute.

Zlitni, Sami et Liénard, Fabien, 2012 (dir.), *La communication électronique dans la " société de l'information ". Quels usages ? Quelles pratiques ?*, Mont Saint-Aignan, Editions Klog.

Zlitni, Sami et Liénard, Fabien, 2013a, "La communication électronique comme élément de la trace numérique", in NETCOM (Networks and Communication Studies), n° 26, 1/2, Retrieved from <http://www.netcom-journal.com/>.

Zlitni, Sami et Liénard, Fabien, 2013b (ed.), *La communication électronique en questions*, Berne, Peter Lang.

Yves Winkin

Prolegomena. "Do we really need to fly from communicational utopia?"[1]

Liminary

Good morning everyone, and thank you for your kind hospitality. Fabien Liénard thanked me (profusely) for coming over, but that should be the other way round. I should thank you all for bearing with me the time of a whole keynote address while I am not a member of your community of experts on "electronic communication". As you may know, I am on the "interpersonal" side of communication. I called it years ago the "anthropology of communication" in order to carve a niche for myself in the field. And I have always been uncomfortable with "mediated" communication, "liquid medias", "cyberculture", and the like. So when Fabien Liénard asked me to come over, I delivered a rather obfuscated title and an even more abstruse abstract, in order to let myself provisionally off the hook. But my idea was the following: one way or another, the anthropology of communication as I suggested it has to deal with electronic communication; it cannot just walk away from it. Either a dialogue, a partnership, a joint ground is established with you, as experts in the field, or the anthropology of communication is going to die out pretty soon, just like many other conceptual dinosaurs which didn't keep up with evolution (I am willing to accept, by the way, that many such conceptual dinosaurs died out of the stinch of their farts, as the kids' joke goes).

So my program is the following.

I am first going to offer you a personal history of the anthropology of communication, and tell you about my shy attempts at dealing with the usages of Internet in the late 90's.

1 It is a text of a plenary conference presented by the professor at the international conference "Electronic communication, culture and identities" which took place at the Information-Communication Department of the IUT of Le Havre in June, 2014.

I am then going to suggest a bridge between the anthropology of communication and the development of "commons" in recent years. After all, "communication" is still a term of many surprises.

And finally I will unravel the mysteries of my title and abstract. I wouldn't like you to be too frustrated when you rush for coffee at 10:30.

A personal history of the anthropology of communication

1967: the inaugural paper

In 1967, the anthropologist and linguist Dell Hymes, published a paper titled "The Anthropology of Communication" in a book edited by Frank Dance under the title *Human Communication Theory: Original Essays*.[2]

Those were the days when *communication* was much discussed as a central concept for the social sciences. Shannon and Weaver's mathematical theory of communication (1949) had quickly become the dominant paradigm of the 50's and 60's. It was embraced in many disciplines, from linguistics (Jakobson) to anthropology (Lévi-Straus)–and, sure enough, in Communication, as soon as it emerged as an academic discipline.

But there were alternative views, expressed by people like Ruesch and Bateson[3]; Pittenger, Hockett and Danehy[4]; Birdwhistell and Scheflen[5]. Those various scholars, who came from linguistics, anthropology, but also, quite remarquably, from psychiatry, argued that communication could not be seen as an intentional transmission of verbal messages between two persons but as a much broader phenomenon in which all members of a given society participated throughout their life. A summary slogan was: "one cannot not communicate". Their *patron saint* was the Edward Sapir of the "Communication" entry of the *Encyclopedia of the Social Sciences*

2 Frank E. X. Dance, 1967, ed., *Human Communication Theory : Original Essays*, New York, Holt, Rinehart and Winston, pp. 1–39.

3 *Communication : the Social Matrix of Psychiatry* (1951).

4 *The First Five Minutes* (1960).

5 Their ideas were expressed in various papers published in the 60's and offered in books in the early 70's (*Kinesics and Context*, 1970 ; *Communication Structure : Analysis of a Psychotherapy Transaction*, 1973).

(1931): "every cultural pattern and every single act of social behavior involves communication in either an explicit or an implicit sense". One of their allies was the Margaret Mead of *Continuities in Cultural Evolution* (1964), for whom there were communicative acts as short as an interaction and as long as a generation. But overall, those scholars were still voices in the desert; their vision of communication was counter-intuitive, and thus hard to grasp. Shannon and Weaver's model was so visually vivid; it fitted so well with the popular representation of mass media, especially television, that it was difficult not to have it take over about every textbook in the country.

Now comes Dell Hymes's paper, which is a call for a new discipline, under the aegis of anthropology. In a very astute way, he is going to reject both the technical model of communication (Shannon and Weaver) and the all-encompassing model (Bateson): "From too narrow a conception of communication, the pendulum may swing to too broad a one" (p. 17). His criticism against the definition of communication as the whole culture is based on a maxim attributed to Birdwhistell: "Nothing never happens". For Hymes, the maxim reflects the point of an external observer, but not the point of view of a participant within a culture, for whom it may well be that nothing happens in a given situation[6].

Against too narrow a perspective on communication, Hymes is going to argue in favor of intrapersonal communication, like prayers, dreams, and visions, so long as they are publicly reported afterwards. This is quite a bold step against a strictly intentionalistic definition of communication[7].

But Hymes is going to go one step further, as he is going to consider "attributed intentionality" in the communicative process, ie the sender may not be "human": he may be a god, a dead person, an animal sending messages to the receiver, who attributes an intention to communicate to the sender. Hymes uses a well known example, offered by Hallowell in

---

6 "To an external observer, what goes on in an interval not considered communicative by its participants may be communicative indeed, but he constitutes it so within a system one degree more complex than that which it is the business of ethnography, in the first instance, to disclose" (p. 20).

7 "Most intrapersonal communication will not be directly accessible to the ethnographer, but reports of much of it will be. Here, of course, personal experience, including spiritual encounters and dreams, shade into public communicative events sich as myth recitals. Cultural stylisations of accounts of private communications, and cultural criteria for accepting such individual reports as valid, are of the greatest importance for study" (p. 23).

"Ojibwa ontology" (1963)[8]. In the traditional Ojibwa's "communicative economy", to use a term Hymes coined in another paper, gods speak to humans through thunder. It is just perfectly normal, "natural". Any anthropology of Ojibwa communication would include thunder as a communicative device. Hymes concluded:

> The scope of "communication" in anthropology must thus depend upon the scope of communication in the individual cultures, or communities, on whose ethnographic study anthropological fact and theory rest. In any given culture or community, behavior, and objects as the products (by makeup and location) of behavior, are selectively organized, used, attended to, and interpreted as communicative (p. 25).

Hymes' anthropology of communication was based on the "emic" perspective that any ethnographer must adopt as his *raison d'être*. If the members of a community consider that cats and dogs talk to them, then cats and dogs are communicative partners, and must be studied as such by the anthropologist. Hymes opened up the realm of communication to extents not even considered by Bateson, Birdwhistell and the other champions of an all-encompassing view of communication. It was a bold, imaginative and "exciting" approach to communication. But it was not received as such, so far I was able to figure out. The paper is barely known, and even less mentioned.

Ten Years Later

In 1976, I was admitted in the Master's program of the Annenberg School of Communication. As an undergraduate student in Communication at the

8 "An informant told me that many years before he was sitting in a tent one summer afternoon during a storm, together with an old man and his wife. There was one clap of thunder after another. Suddenly the old man turned to his wife and asked, 'Did you hear what was said ?' 'No', she replied, 'I didn't catch it'. My informant, an acculturated Indian, told me he did not at first know what the old man and his wife referred to. It was, of course, the thunder. The old man thought that one of the Thunder Birds had said something to him. He was reacting to this sound in the same way as he would respond to a human being, whose words he did not understand. The casualness of the remark and even the trivial character of the anecdote, demonstrates the psychological depth of the 'social relations' with other-than-human beings that becomes explicit in the behavior of the Ojibwa as a consequence of the cognitive set induced by their culture" (pp. 24–25).

University of Liège, I conducted interviews with book publishers in order to write a memoir on book publishing in the French-speaking community of Belgium. I was struck by their "impression management" strategies. I became so fascinated by Goffman's *Presentation of Self in Everyday Life,* that I decided to come to Penn for graduate work.

I was going to discover a rather extraordinary intellectual environment (well described by Leeds-Hurwitz and Sigman, 2010) of which Goffman was an element, but certainly not the only "star". I took Dell Hymes' course on the "ethnography of speaking". He was not, pedagogically speaking, a captivating professor, but his ideas were pure gems to me. The well balanced combination of pedagogical and intellectual brilliance was offered by Ray Birdwhistell. Birdwhistell was a tall, seductive man, with a deep voice and a huge smile. He would pace in front of the audience with odd questions like: "Yves, tell me how you start cleaning up under the shower? " I still remember my embarrassment in front of my schoolmates, who were just too happy to have been spared the same question. Birdwhistell's idea was to lead to the notions of embodied knowledge, cultural embodiment, etc. *Balinese Character*, Mead and Bateson's masterpiece (1942) was one of the required readings. We would have to read several books among the monuments of the field. No recent literature. But through that combination of readings and personal anecdotes about the giants of the field he used to know quite intimately like Mead, Bateson, Warner, we would acquire a sense of the history of the field, and a sense we were part of that history: we were going to write the next chapter of it.

Birdwhistell barely ever told us about kinesics, the linguistically based approach to body motion communication he created in the 50's, but he had us read his papers on "multi-channeled communication". We all learned "We don't communicate; we participate to communication"; "There is no more 's' at 'communication' than there is at 'gravitation' " – and "nothing never happens", sure enough. He supervised my MA thesis on the dynamics of intercultural communication at the International House of Philadelphia, and he corrected line after line of the manuscript, always asking me to be more precise in my descriptions: "Dim light is Winkin's" was one his remarks that I remember to this day. He was above all an acute observer of "small behaviors", fascinated by their choreography, like in his famous paper on " The Cigarette Scene"[9].

9 Sapir's well known sentence about gestures was his primary source of inspiration: "Gestures are hard to clarify and it is difficult to make a conscious separation between

La Nouvelle Communication

When I returned to Belgium in 1978 to become a fellow of the Belgian National Science Foundation, I was decided to produce a book around Birdwhistell's ideas on communication. I didn't know anybody in Belgium or France who may be interested, except for… Julia Kristeva, who had briefly written about kinesics in a linguistics journal (*Langages*). So I wrote to her–and she responded, to my amazement. She refered me to François Wahl, an editor at her publishing house, Les Editions du Seuil. When I called to make an appointment, Wahl was not available, and I was referred to his young assistant, Jean-Luc Giribone. I am telling the anecdote to show you how happenstantial a career may be. Giribone was about my age, and he was just back from a two-year teaching bout at Yale, as a graduate of the Ecole Normale Supérieure. He was in charge of introducing the so called "Palo Alto School" to the French public through translations of Bateson's, Watzlawick's and others. He had not edited many books yet. He definitely wanted to impress his boss, François Wahl, who was the editor, I discovered later, of Roland Barthes, Michel Foucault and other French luminaries. My project fitted Giribone's plans, so long as I agreed to include the Palo Alto School next to the "Philadelphia School", as I called it (Birdwhistell, Scheflen, and a representative of the next generation, Stuart Sigman). And I had to deliver the manuscript before Christmas 1980. I had about a year.

I managed to deliver it on time (everyone at home was involved in the frantic typing of the manuscript). And it got published in May 1981 under the title "La Nouvelle Communication", I have to admit that the election of François Mitterrand as President of the French Republic that month overshadowed the event. But to my surprise, sales started to pick up in the Fall. Sales haven't stopped since.

---

that in gesture which is of merely individual origin and that which is referable to the habits of the group as a whole. In spite of these difficulties of conscious analysis, we respond to gestures with an extreme alertness and, one might almost say, *in accordance with an elaborate and secret code that is written nowhere, known by none, and understood by all*" (p. 556, italics are mine).

Anthropologie de la communication: de la théorie au terrain

For years I wanted to write a sequel to *La Nouvelle Communication*, which would explain how to go about doing research within the theoretical frame provided by the "orchestral perspective". For reasons which are probably close to the "second book syndrome" developed by many authors whose first book is too successful, it took me 15 years (1996) to finally publish a provisional book titled *Anthropology of Communication: from theory to the field*. A definite version was published in 2001 by Le Seuil in a pocket version.

What I basically developed was a call for ethnographic inquiries based on Bateson's and Birdwhistell's wide open definitions of communication. But I stressed that Hymes' "attributed intentionality" was a crucial addition to the "platform", while Goffman's work, no matter what he said about communication, was to be injected as well, because it fitted quite nicely. In the background, there were references to Sapir, but also to Ward Goodenough, whose definition of culture was perfect for my project :

"As I see it, a society's culture consists of whatever it is one has to know or believe in order to operate in a manner acceptable to its members, and to do so in any role that they accept for any one of themselves" (1956).

I defined communication as "the performance of culture", both in reference to Goodenough and to Goffman, whose *Presentation of Self* was still quite seminal to me. And so it went. The second part of the book was devoted to fieldwork methodology , and the third part to five brief pieces of fieldwork I had conducted in different contexts: as a student at the International House of Philadelphia, as a tourist in Marocco, as a visiting scholar at an American university. The message was aimed at students: you can do it as well; you don't need a two-year immersion in a farfetched place like aspiring anthropologists had to do. Ethnography needs not be a gruelling rite of passage. It can be fun – so long as you have a good conceptual frame to help you avoid turning into a mere story-teller.

The anthropology of communication as I conceived it was not focused on specific objets of inquiry. I saw it as an "analytic frame" ready to give shape to inquiries in many domains. Rather than an anthropology *of* communication, which seems to imply that "communication" is the object of inquiry, I suggested that an "anthropology *through* communication" (or "*from* communication") was in order (in French you can play with the "de" as in "depuis"–with a reference to the Latin *ab*). In that respect, I was departing from Hymes's paper but I was still under the spell of his master, Edward Sapir, who wrote in 1927:

"Strange as it may seem at first blush, there is no hard and fast line of division as to class of behavior between a given style of breathing, *provided that it be socially interpreted,* and a religious doctrine or a form of political administration."[10]

In other words, I believed and I still believe that communication must remain a theory-grounded interpretation, not a collection of objects, and certainly not of technology-based objects.

Let's sum up.

If Sapir, Mead, Bateson can be considered as the first generation of scholars related to an anthropology of communication, Hymes, Birdwhistell, Goffman would the second generation. Sigman, Leeds-Hurwitz and myself would the third, while Emmanuelle Lallement, Véronique Servais, Filareti Kotsi (to name three scholars in their 40's who worked under my supervision) would be the fourth. And now comes the fifth generation, constituted by digital natives, who are going to challenge the physically situated interactions of the former generations. So let me address the question that must tackled for a future to be at all.

Anthropology of communication and "liquid media"

Let me start with my own experience. In a book titled *Comment l'informatique vient aux enfants. Pour une approche anthropologique des usages de l'ordinateur à l'école (*2006), Eric Barchechath, Rossella Magli and myself report on a study we conducted in the mid 90's in different European countries to observe how 5th graders learn to use computers. We focused on their interactions with the teacher, with the machines, among themselves and with their peers from other schools, as they contacted them through e-mails. The book had one major flaw and two good leads.

The major flaw had to do with "misplaced concreteness". To paraphrase Katerine Wahl-Jorgensen, it was "classroom-centric", since ethnographers love well-bounded situations. After all, anthropology was born on islands (Trobriands, etc). The real challenge for an anthropology of communication is to deal with the "liquidity" of Internet and the new

10 Sapir, Edward, 1927, "The Unconscious Patterning of Behavior in Society", in Dummer, E.S., ed., *The Unconscious : A Symposium*, New York, Knopf, pp. 114–142.

media. In other words, the anthropology of communication has to integrate "digital ethnography" or "net ethnography"[11] in its methodological tool kit.

Let me make a brief detour through a paper by Joel Sherzer, one of Dell Hymes's early students. In 1978, Joel Sherzer critically assessed Hymes' efforts to get an "emic" sociolinguistics off the ground[12]. One of his suggestions was to train young natives to study their own community as local participant observers (p. 52–53). I would suggest that those young natives are already at work as far as cybercultures are concerned. Young communication scholars of the "fifth generation" are "digital natives"; they are immersed in cybercultures since they were born. A reflexive auto-ethnographic attitude could lead them to produce thick descriptions of the practices of their digital community or communities. Immersion in a virtual field is a most difficult operation for an anthropologist (of communication or not) who had no previous first-hand experience with ICT's. But for those young anthropologists who lived for the entire length of their life in intimate contact with them, immersion is not a problem. Look at the way Michael Wesch and his students entered YouTube as "digital ethnographers" in 2007. They were among the first to show how YouTube was questioning old-established notions of public space, presentation of self or social identity. Sure Wesh could not escape some clichés about the "loss of community" and "people connecting very deeply ", but he was exploring. Michael Strangelove[13] and others came just after him.

So the challenges presented by Internet and the digital media to the anthropology of communication are not so much methodological than conceptual. Is the conceptual platform of the anthropology of communication, derived from the work of scholars who certainly were no digital natives, obsolete? Sure, I am going to say no for two reasons:

First, when one peruses through a up to date book like *The New Media and Cybercultures Anthology* edited by Pramod K. Nayar (2010), one

11 Wittel, Andreas, 2000, "Ethnography on the Move: From Field to Net to Internet [23 paragraphs]", in *Forum Qualitative Sozialforschung / Forum: Qualitative Social Research*, *1*(1), Art. 21. Retrieved from <http://nbn-resolving.de/urn:nbn:de:0114-fqs0001213>.

12 Sherzer, Joël, 1978, "The Ethnography of Speaking : A Critical Appraisal", in Saville-Troike, M., ed., *Linguistics and Anthropology. Georgetown U. Round Table on Languages and Linguistics 1977*, Washington, D.C., Georgetown University Press, pp. 43–57.

13 Strangelove, Michael, 2010, *Watching YouTube*, Toronto, University of Toronto Press.

can certainly see that most authors cited are not digital natives. When you think about it, if authors have to be of the same age as their objects, we are going to be in trouble in philosophy and quite a few other disciplines in the humanities and the social sciences.

The second reason to say no is more substantial. Let me return to my study of children discovering Internet.

The project involved collecting a long story built by several classes of different Italian schools of Emily-Romagna involved in a ring, so to speak. Each class added a chapter. The story had to do with kids sucked into the computer at the moment of closure. They got trapped in a big cavern, inside the computer, and they were chased by creatures called "Sbirulis", who lived inside the computer. They were the "bugs" their teachers had told them so much about (remember, we are in the mid-90's, and there were plenty of bugs in those days). The kids finally escaped when someone opened the computer again. What we found fascinating was the kids' capacity to believe and not believe at the same time the story they created. That's what I call "enchantment": a capacity to willingly suspend rational judgment, a capacity to "live elsewhere" for a while.

Enchantment again

A few of you who know my work may now be thinking: "here he goes again". Yes, I dare suggest here that anthropology of communication addressing Internet cultures should consider the notion of enchantment as a analytical tool. Let me explain. For many years, I have been trying to explore social situations in which the famous Coleridge's formula "the willing suspension of disbelief" applies. I have been arguing that in many social worlds (trips abroad, PR events, residential conferences, group dynamics seminars, carnivals, cruises – and of course Disneylands and other theme parks), a collusion between "enchantment engineers" and participants lead to a temporary "suspension of disbelief", provided that participants keep denying the reality around them with a mantra like "I know, but").

The general hypothesis I would like to offer is that Internet offers, not only to schoolchildren, but to users of all ages, the possibility of an enchantment, ie the possibility to willingly suspend disbelief while you momentarily dwelve into a fantasy world. Take the "society of friends" that

Facebook invites you to believe in; take "Second Life" and Groland; take Twitter and the pseudo-egalitarian sharing of bits of news and comments.

I know: I am here on Sherry Turkle's turf. You know her brilliant books: *The Second Self* (1984), *Life on the Screen* (1995) or the most recent *Alone Together* (2011). While she doesn't speak of enchantment, she comes close when she describes how students got involved in Weizenbaum's ELIZA, "a program that engaged in dialogue in the style of a psychotherapist" (Turkle, 2011: 23):

> Yes, I thought, they (the students) engaged in personal personal conversations with ELIZA but in a spirit of "as if". They spoke as if someone were listening but knew they were their own audience. They knew all about ELIZA's limitations, but they were eager to "fill in the blanks". I came to think of this human complicity in a digital fantasy as the "Eliza effect" (p. 24).

Turkle is going to explore the many facets of the "ELIZA effect" through the book: how we deal with machines "as if"; we know they are only machines – but. Although the ELIZA effect is not the exact equivalent of "my" enchantment, there are overlaps and there is room for further ethnographic – and conceptual – exploration. What I am suggesting is that "enchantement"is a "sensitizing concept", as Herbert Blumer[14] put it: it is like a magnet, it attracts data and helps figuring patterns. But it is not an explanatory concept yet. It needs further elaboration.

E-mails as gifts to be received and returned

Let me return one last time to my study of schoolkids exploring the possibilities of Internet. We also discovered that kids had to be told and reminded several times to respond to the messages they had received. Instead of framing that lack of responsiveness in psychological or technical terms (eg. the kids are not too comfortable with the keyboard, they don't know what to say, they are uncertain of the spelling, etc), we used a sociological explanation: kids have to be socialized in terms of "giving, receiving and reciprocating". Like all the generations before and after, children have to be trained to deal with gifts, following Marcel Mauss. It is a most fundamental aspect of social life. They have to learn how to say "thank you",

14 Blumer, Herbert, 1954, "What is Wrong With Social Theory?", in *American Sociological Review*, Vol. 18, pp. 3–10.

they have to learn to reciprocate (not too fast, not too slowly), they have to learn to offer. For the current generation, e-mailing is one of the major ways to learn those rules. Kids learn that gifts (mails) have to be returned to their senders within an appropriate lapse if they wish to be considered as good partners in life.

What I am saying here is that old (but sound) sociological ideas like gift-giving can give a profound dimension to data collected about the uses of digital media. They can recast cybercultures within the history of processes of socialization from one generation to the next. Reversely, data on new media can give a fresh life to old concepts. I am thinking here of "community" which is certainly returned to active service since Howard Rheingold's *Virtual Community* (1993). I am also thinking of "interconnectedness", which was already used by Lloyd Warner in the 40's, and recast by one of his students, Ray Birdwhistell, to define "communication as a processual point of view on interconnectedness"[15]. And finally "communication" itself can both refuel the analysis of virtual spaces and be refueled by it.

The anthropology of communication and the emergence of commons

As we all know, *communication* and *communion* were synonyms for many centuries. It is not until the 17th cy, that communication became a synonym of transmission. But the old meaning of sharing is still embedded in the term: *munus* (com-munus) was used in Latin for the "public show" that recently elected public officials would give to their voters. *Munus* is both a present and an obligation. Marcel Mauss again. At the heart of communication, there is the *mana* of the gift…

It is actually a line of approach used by observers of *commons*, who assemble within the same social category the age-old collective maintenance of water systems in traditional societies, like 2010 Nobel Prize of Economics Elinor Ostrom studied, and "creative commons" like Wikipedia,

15 Birdwhistell, Ray, 1970, *Kinesics and Context*, Philadelphia, University of Pennsylvania Press, p. 251.

Blender movies or "crowd" initiatives[16]. What I find pretty fascinating is that the very term *commoner* can be translated by the old French term *communier*, still available in the Littré dictionary. A "communier" is a farmer who share his piece of land with other farmers, like today commoners are sharing platforms for creative purposes. What I find pretty fascinating is that a link is reinstalled between medieval practices and "avant-garde" practices. As if Margaret Mead's suggestion about communicative acts which takes many generations to get accomplished was fully demonstrated. As if the definition of communication as "performance of culture" was taking a new meaning: not so much a performance of culture in the interactional here-and-now ("the punctiform present", as Birdwhistell used to say) but a performance of culture in a much more elongated time frame, stretching over several generations. In other words, I have a hunch that an anthropology of communication may be useful to study commons as they keep mushrooming all over the world today.

The title finally

In 1992, the French sociologist Philippe Breton published a very short and sharp book, titled *L'utopie de la communication.*[17] He was trying to answer the question: how come "communication" has become so popular is so many domains? His suggestion was that communication had become the ultimate utopia. But he warned against the perverse effects of such an utopia. His argument was that the origins of modern communication was to be found in Norbert Wiener's book, *Cybernetics and Society. The Human Use of Human Beings* (1948). But, according to Breton, Wiener was dreaming of a transparent society that would never be able to produce again a Holocaust. Holocaust was only possible because of the capacity of a society for deep secrets. The utopia of communication is thus a transparent society, says Breton, but it is also a potentially totalitarian society:

16 I am using here David Bollier's book, *La Renaissance des communs*, Paris, Charles Leopold Meyer Editions, 2014, as prefaced by Hervé Le Crosnier.

17 Breton, Philippe, 1992, *L'Utopie de la communication*, Paris, La Découverte.

> Avec la communication a resurgi le mythe d'un thème vital unique, qui traverserait toutes les activités humaines en les englobant. Le présupposé majeur de ce nouveau paradigme est le lieu commun selon lequel "tout est communication" ou, sous une forme plus acceptable, qu' "il y a de la communication partout". Il suffirait dès lors de trouver les lois générales de fonctionnement de la communication pour bâtir une conception unitaire du monde et, surtout, disposer des moyens de le maîtriser.[18] (p. 120)

So do we necessarily have to fly away from communicational utopia? Breton would like us to stay away from the utopia of communication he analyzed. But I would argue that we may stay close to the new communicational utopias offered by the current "commons". Not only they offer grist for our mill as social scientists, but they offer some glee in these times of morosity. Commons enact in new ways the old meaning embedded in the term "communication": gift/counter-gift, sharing, trust. Both as a social scientist and a citizen, I feel like there is some kind of a renaissance looming ahead. There is still a future for the anthropology of communication, and there is still a future for my son. That's pretty comforting. Thank you.

18 Philippe Breton, *op.cit.*, p. 120.

ICT and political communication

Pitabas Pradhan

Social Media Impact on Election Campaign: A Study of the Indian General Elections 2014

Introduction

Although various election studies have suggested that media do not change voter's choice and merely reinforce the existing attitudes and believes, political parties leave no medium unturned to influence voters during election campaigns. Impression prevails that leaders rise to great heights through image created by media. The outcome of the Indian general elections 2014 has formed the impression that pro-active campaign strategy has overpowered core factors of voting behavior such as casteism, regionalism, religion, money, ideology, opinion leaders etc. in influencing voters. Election has virtually becoming a communication contest, political parties employ various communication tools to mobilize the voters.

With growing internet penetration in India social media has become an important tool in election campaign. The emergence of interactive Web 2.0 applications such as blogs, micro blogs, social networks, and photo and video sharing sites have resulted in a paradigm shift in political communication. The social media tools have opened up new avenues for self-expression and information sharing that can bring alternative voices into the public sphere.

The new tools enable anyone to research, create and disseminate content, extending news publishing to the street. By promoting online discussions, it has been strengthening a conversational democracy, where "citizens and political leaders interact in new and exciting ways". The rapid expansion of mobile networks in developing economies like India has taken the Internet to the palms of the people in the remotest of the countryside and made the impacts far more visible.

Despite political communication being influenced by the established socio-cultural values, ICTs are innovatively used to carry the political messages across to the people in one-to-many and many-to-many modes catching the imagination of the common man. ICT based political communication campaigns in India have altered the dominant political alignments giving way to new political identities and larger than life images of certain politicians. The Delhi Assembly elections 2013, and the national elections 2014 were testimony to the growing influence of ICTs in political communication in the country. The unparalleled rise of Narendra Modi, and AAP chief Aravind Kejriwal speak lot about the role of social media.

The attempts to organize crash courses on use of social media by major political parties, and the Election Commission of India's guidelines on the use of social media during elections might be based on the perception that in the present age of online communication, election campaign cannot afford to ignore social media. The result is the politicians and parties are taking to Facebook and Twitter in a big way. In Delhi, an urban state with a literacy rate of over 86 per cent, and a per-capita income of three times that of the national average, social media is expected to have a bigger role. The situations in other metro cities were not much different. The rural areas with low literacy, connectivity and per-capita income are expected to have limited social media influence.

Against this backdrop, the present study seeks to explore the role of social media in the Delhi assembly elections 2013 and the Indian general elections 2014. The two consecutive elections have been included within the purview of the study because much of the campaigns started concurrently. The specific questions this study attempts to answer are:

How important was social media in the overall media plan of the mainstream political parties?

To what extent the parties made use of social media components?

How the mainstream parties integrated social media elements into their election campaigning strategies?

Have the social media tools influenced the impacts of political communication?

Have social media influenced voters and hence electoral outcomes?

Conceptual Framework of the study

The Merriam Webster's dictionary defines social media "as the forms of electronic communication (as Web sites for social networking and micro blogging) through which users create online communities to share information, ideas, personal messages, and other content (as videos)"[1]. Andreas Kaplan and Michael Haenlein define social media as "a group of Internet-based applications that build on the ideological and technological foundations of Web 2.0 and that allow the creation and exchange of user-generated content. At the core of social media is the social interaction among people in which they create, share or exchange information and ideas in virtual communities and networks. There is a wide variety of social media, ranging from social sharing sites such as YouTube and Flickr through social networks such as Facebook and LinkedIn. Social media differ from traditional in many ways, including quality, reach, frequency, usability, immediacy, and permanence. For content contributors, the benefits of participating in social media have gone beyond simply social sharing to building reputation and bringing in career opportunities and monetary income (Tang, Gu, and Whinston, 2012).

Social media has some unique user-friendly features, which may be attributed to their spectacular popularity.

First, social media operate through an individual's social network. Information flow through social media networks vis. friends and friends of friends.

Second, the social media networks function at very low cost to the user-cost of digital service and access to the network.

Third, social media networks have multiple channels of access- can be accessed on cell phone, computers, laptops, or any other digital device available in multitude of forms today.

Fourth, social media networks are easy to use- they require no specialised technical knowledge beyond basic computer literacy and feature platforms that can be mastered in minutes.

Advent of social media and its impact

As the information communication technology (ICT) advances presented new tools to enable anyone to research, create and disseminate content, the job of publishing extended from the newsroom to the street. Publishing became possible from anywhere beyond the confines of institutionalised media newsrooms. The growing smartness among the social media designers, made it more focused and personalised through tagging, lists, and other filters enabling meaningful communication and instant participation. It empowered the individual who had a story to tell with the tools to proclaim their news to the world.

The traditional media audience moved away and found a new home in their preferred social networking space, where they met like-minded individuals and formed communities (Friedrichsen, 2013)[1]. Based on the quality of the contents they produced, people began to gain followers through peer group recommendations. As these networks expanded, bloggers and social networks emerged.

These content producers are individuals who wrote and broadcast with authority, not because they were paid to do so, but because they had knowledge and passion that they wanted to share with the world. Gradually, networks began to be built; experts linking their skills and sharing information. For many, social media is a central part of the news gathering, news production and news distribution strategy. The content they produce contains audience inputs which the audience thrusts and, because it discusses issues that concern them, they comment, add value and share.

Impact on traditional media

Alert traditional media organisations formulated social media strategy taking note of changing audience behavior. As was evident from the Delhi gang rape case of Dec 2012, traditional media have been adapting to the social media. The social media flash of the news, forced the traditional media to further narrow down the event-telecast time, many of the norms that had guided old media were discarded. Unmediated broadcasting of

1 Mike Friedrichsen and Wolfgang Muhl-Benninghaus, 2013, *Handbook of social media management*, Berlin: Springer, P.3.

viewpoints straight from the ground in real time got a boost (Philiphose, 2013:20).

Social media and democratization

By promoting online discussions, social media has been strengthening a conversational democracy, where "citizens and political leaders interact in new and exciting ways". A.J. Leibling claims that: "Freedom of the press belongs to those who own one." But, the Internet enables everyone who owns a computer to have his or her own printing press. Arianna Huffington, a well accepted figure in blogosphere, claims that with blogging self expression has become the new entertainment" and people are increasingly likely to participate online than just vegetate in front of the television. Jeff Bullas claims that "blogging has been a path not only to self expression but self fulfilment that has become a journey of self discovery". The social networking websites allow individuals to create a personal profile and form connections with other individuals with whom they can relate, and share current and future experiences. Huffington says "I write about topics that resonate with my soul and purpose while keeping focused and relevant. It is an exciting time to live where you can connect to your global tribe with rich multimedia that transcends time and place".

Social Media and the rise of participatory culture

"In confronting the challenges of participatory culture: media education for the 21st century", Henry Jenkins explains that the rising participatory culture is characterised by "relatively low barriers to artistic expression and civic engagement, strong support for creating and sharing one's creations, and some type of information mentorship whereby what is known by the most experienced is passed along to novices" (Jenkins, 7).

Being active engaged in participatory networks, people of all ages and back ground create and distribute messages via online and interpersonal networks at a rapid and ever accelerating speed. Armed with inexpensive tools of capturing, editing and organising, people tap into a vast ocean of real-time data and multimedia content to promote personal and political interests.

Functions once monopolised by handful of hierarchical media institutions have been usurped by independent publishers, video sharing sites, collaboratively sustained knowledge banks, and fan-generated entertainment.

Social media and the rise of digital activism

Social media tools are used to reveal corruption, organize protests, mobilize public support for a cause. Digital participations like Facebook groups, blogs, and text – messaging, just to name a few, fundamentally alter certain types of interaction among the public and provide individuals with avenues to participate in the public life. Digital activists like Wael Abbas creator of the well-known website "Misr Digital" used digital technologies to start a protest wave that swept the Arab world overthrowing the authoritarian government. In 2011, hundreds of thousands of Egyptians responded to a call to protest which began on face book eventually leading to the fall of the Hosni Mubarak regime. During the civil society movement for a Lokpal in India, the miniature multitasking devices, the social networking sites and blogs having galvanized millions throughout the length and breadth of the country pushing them to the streets in support is a testimony to the social media's power of public mobilization.

Social Media Usage in India

According to social media data released by 'we are social' in January 2014, there were 213 million internet users in India with penetration among 17 % of the population.

The average number of hours spent on Internet through Laptop/ Desktop is 4.9 hours.

Whereas the average time spent of Internet through mobile is 2.6 hours. Mobile internet penetration was among 11 % of the population. There were 90 million social media users in the country with a social media penetration of 12 %. Average time spent on social media being 2.4 hours.

Active mobile social users stand at 72 millions. There were 893.86 million mobile subscribers with a penetration of 73 %. Seventy five per cent of the active Internet users in India use social networking sites like Facebook, Twitter, YouTube, Pinterest etc.

Table 1. Social media use in India.

Network	Ownership (% of users)	Used during last month (% of users)
Any social network	97	72
Facebook	94	55
Google+	78	35
Twitter	67	30
LinkedIn	54	24
Orkut	51	17

Source: <http://was-sg.wascdn.net/wp-content/uploads/2014/01/slide092.png>, retrieved Sept. 15, 2014.

India will account for the third-largest user base on micro-blogging site Twitter at 18.1 million by the end of this year surpassing UK according to research firm eMarketer reports. It will see the most consistent growth pattern.

According to Social Media in India report 2013 released by the Internet and Mobile Association of India (IAMAI) and IMRB, social media users in Urban India in October 2013 was 86 million. The report further says that 19.8 million people in urban India access the social media platform through mobile phone. It also says that young male population are the highest users of social media contributing to 30 % and 26 % users in the top four metros and non-metros respectively. Interestingly, the report also says that younger women are increasingly using social media, whereas, in the Other Metros & Small Metros, "College Going Students" register the highest proportion of social media usage.

Social Media as campaign tool in Indian general election 2014

After the Delhi Assembly elections 2013, the 16th general elections 2014 was the first national political campaign in which political parties in India attempted to woo voters on social media. The influence of social media was so huge that analysts call it India's first "social media elections"(Kugelman). It also speaks volumes about the election campaigns of the future.

Politicians across party line took part in Google+ Hangouts, televised interviews organised by Facebook, online political debates, and widely used WhatsApp to connect with millions of tech-savvy urban voters. Parties were also interacting with voters on their mobile devices. WhatsApp was being used by all major parties including BJP and Congress to send photos, videos and messages to potential voters. In its membership drive campaign the AAP members urged people to give them a missed call for free, party officials then got in touch and formally enrolled them as supporters.

"Now no serious politician is seen as being able to avoid social media altogether" claims Congress leader and former union Minister Sashi Tharoor, who happens to be a smart social media user. "I've seen social media become more important," says Anshul Tewari, founder of *Youth Ki Awaaz*, or Voice of the Youth, a popular blog that screens Google Hangout sessions with leading politicians where readers can send in questions. 75 percent of India's online population was younger than the age of 35, speaks volumes about the social media role in political communication. Politicians realized that social media is not only a couple of kids talking about fashion, but also people talking about serious things.

As the election campaign reached a frenzied peak, social media became the battleground for the major political parties. Personalised greetings by politicians to followers on twitter, celebratory recorded phone messages that concluded with an appeal to vote became common. Politicians, with their traditional kurtas and caps increasingly turned to social networking sites to engage Internet-savvy first time voters. As the social phenomena spread on the cyber space, having an official YouTube channel or an active Facebook page seemed to be as important as holding mass rallies and displaying huge billboards.

Political parties were using tools to crunch the insurmountable amounts of information social media generates. Most parties hired services of expert PR agencies and online platforms to help them make content viral on the Web. Pinstorm, a digital marketing agency engaged in collecting and analysing political statements over online platforms was used by AAP to compare how the party was faring against others.

Parties introduced innovative low-cost strategies for the Indian environment.

Campaign groups have sent volunteers out to remote areas with solar-powered projectors stored in backpacks, so campaign messages could be shown directly to voters. Similarly, in areas where on-the-ground campaigning and registration drives were needed, websites, email and Google Hangouts were being used to co-ordinate volunteers.

Seeing the Indian youth enthusiastically embracing social media as their first tool when they started using the internet, political parties ensured their presence on the online platforms. "Standard tools like Facebook, YouTube, Twitter and then lately WhatsApp, were the pillars of our strategy," claims Arvind Gupta, head of BJP's IT and social media cell. One of the most interesting uses of social media was when the BJP crowdsourced its manifesto, with hundreds of thousands of people using Twitter and blogs to comment. "A lot of those suggestions were actually incorporated in our manifesto," Gupta said. The BJP used a two-pronged approach to social media: increase its online presence while also helping in its offline activities. In a program called "organize online to assist offline," the BJP used the reach of social media to recruit volunteers, eventually enlisting 2.2 million, Gupta said.

During the election campaign Narendra Modi, BJP's prime ministerial candidate reached out through his Twitter account to the youth of the country, appealing to what they were looking for: "jobs, security and the use of technology," and succeeded in gaining millions of followers. He orchestrated a "Chai Pe Charcha" on giant TV screens across tea stalls in the country, where satellite and mobile technologies were used to host live interactions with people. Hired experts were given full liberty to manage their social media coverage.

Other political parties have also turned to Facebook interviews, mobile crowd sourcing, or hashtag wars on Twitter. Arvind Kejriwal and his Aam Aadmi Party was using social media to draw swathe of supporters and maintained a Facebook page for every single constituency in the national capital. A Facebook tracker Showed how much the candidates and parties were talked about in the social media circuits. Over half of the country's 1.2 billion population already having owned one, mobile became integral to the social media strategy. The BJP used it as an electioneering tool to allow potential voters to listen in on Mr Modi's rally speeches in real time on their phones from anywhere in India. The steady growth in number of smart phone users in the country is giving a surge in social media use. No other medium gives such widespread, simultaneous reach as mobile phone does in India today," claims Milind Pathak from One97 Communications, a Delhi-based mobile marketing firm.

The social media campaigns by the BJP, and Aam Admi Party (AAP) helped "define the media narrative." It created "key messages and talking points" that were then picked up by the mainstream media. Modi was more in tune with young Indians, which the Congress failed to match.

The chief electoral officer's page is updated regularly with videos on the electoral process, photos, rally campaigns, and other information. It attempted to inform people on why greater participation in the electoral process was good for a stronger democracy. Another video explained how an electronic voting machine functions.

Misuse of social media to influence voters was rampant. Delhi Chief Electoral Office Vijay Dev said: Any citizen on behalf of these political parties or candidates can come out with anything on Facebook, Twitter or other social networking sites to influence people and we cannot have any control over that. The election authorities were doing everything to ascertain the content political parties posted and the expenditure involved was put in the shadow registers.

A study by the Center for Media Studies estimates a $5 billion election campaign expenditure by the major political parties of which, according to a report by the Internet and Mobile Association of India (IAMAI), 2 to 5 % is being spent on fuelling their online presence. Twitting by politicians was on the rise. The INC was ahead of others in terms of its wide use of the mobile messaging platform, WhatsApp, both for internal communication and relaying party updates. Google was helping all political parties with their online strategy by organizing weekly hangouts with politicians. The search engine had developed a dedicated election hub, and a "Pledge to Vote" interactive map. "We wanted to engage with first time voters and inspire different age groups to go out and vote," said Sandeep Menon, director of marketing at Google India.

Social media use in Delhi Assembly elections 2013

Population profile of Delhi

Out of a total population of 17,838,842 (2011), 8,976,410 are males and 7,776,825 are females making a sex ratio of 866/1000. The overall literacy rate is 86%. Hindu's constitute 81 percent of the city population, followed by Muslims at 11.7 percent. Among the other religious minorities, Sikhs constitute 5 percent where as Jains and others stand at 1.1% and 1.2% respectively. The city also being the national capital, people from all states of the country stays in good numbers. By virtue of being a great centre of higher learning students comprise a good share of the population. All major political parties having their head quarters, the state is also a nerve centre of political activity. Major national television channels, national newspapers and magazines have their head offices in the city. The country's largest circulated newspapers Viz. the Times of India and the Hindustan Times, having their head quarters her have been engaged in a kind of circulation war. Among others, the national news channels like the government supported Doordarshan, and big private players such as NDTV India, Zee News, ABP News, Aaj Tak, but to name a few, make Delhi a media rich state. Further, the national capital territory has the largest number of centres for media learning.

Social media usage

With penetration over 17% in the metro city of Delhi the role of social media in election campaign was predictable. The accumulation of over hundred thousand university/college student population having access to internet and social media, mostly at the educational institutions make the social media users base much larger. Again, contestants having strong affiliation to higher educational institutions add momentum to social media usage, especially through institutional infrastructure.

A Zee Research Group (ZRG) study of affidavits filed with the Election Commission of India (ECI) by candidates of main parties in the fray for Delhi polls shows that 44,32 and 20 percentage candidates of the three main political parties i.e. the INC, BJP and the AAP candidates have chosen to be away from social media. 31 candidates of the INC, 21 of the BJP and 14 candidates of the AAP are yet to join the social media. 06 candidates of the Aap, 05 of the BJP and 03 of the INC have no e-mail addresses.

Table 2. Online presence of mainstream parties in Delhi.

	E-mail	Facebook	Twitter	Personal website
AAP	65	51	14	00
BJP	66	44	20	14
INC	65	35	09	06
BSP	41	15	01	02

In the present era of global communication networks, party websites play an important element in their media strategy. They provide a centralised forum for campaign information and a key gateway for party workers, supporters, and journalists. But the state-specific websites of the AAP, the BJP and the INC leave much to be desired. Neither the BJP nor the INC has a site customised for mobile phones. BSP does not even have a state specific website. The websites are not updated on a daily basis. The visitors fail to get much inputs on what the parties did or stood for.

However, the heads of campaign and some other prominent leaders, across party lines have many followers on social media. APP chief Arvind Kejriwal has nearly nine lakh and seven lakh plus followers on facebook and twitter respectively. The same party's Manish Sisodia and Shazia Ilmi respectively are also being followed by 41 thousand and 28 thousand netizens. Outgoing Chief Minister Sheila Dikshit is followed by 01 lakh on face book whereas Dr. Harsh Vardhan of the BJP is followed by 15 thousand netizens on twitter. It is to be noted that an IRIS study had identified all constituencies in Delhi as high impact areas where Face book could make a difference between winning and losing.

Campaign strategy

The incumbent Chief Minister of the ruling Indian National Congress (INC), the leader of opposition in the state Assembly of the Bharatiya Janata Party (BJP), and social activist-turned politician leading a new formation Aam Admi party (AAP) were leading the campaign bandwagons of the three major contenders for the state elections scheduled on Dec 04, 2013. The incumbent banked on her development work, the main opposition focused on corruption by the former. The new entrant raised the issue of corruption and raised common men's issues vigorously to rousing welcome by the discontented citizens, especially the younger generation.

The Delhi Assembly elections reflecting the national voting mood in the making, there was, in fact, a virtual war of words among the opposition for getting the space lost by the incumbent government, which was at the receiving end of all abusive language. The opposition parties made optimum use of the media including social media for getting their points across. Series of pre-election opinion polls were, conducted by the parties, were published in the media, claiming that they were far ahead of the other in defeating the incumbent government. This campaign war, most of the times supported with inflated figures, might have confused and influenced many non-committed voters.

It was for the Aam Admi party, the common man who has become cynical of elections, found politics interesting all of a sudden. Banking on the hope of the common man the AAP was hoping to do the impossible of getting majority on its own. With the election symbol of the broom, the Party unequivocally asserted that it is the only honest political outfit, and claimed to take governance to the doorstep of the citizens. The party's campaign to sweep clean the country of corruption has changed it into a movement, fundamentally different from the earlier post independent political movements which overthrew powerful incumbent governments. With an army of volunteers and funds pouring in from both within the country and abroad, the AAP has emerged as a force to bet on. The AAP enthused by its expanding volunteers and swelling coffers through heavy online donations, made full use of advertisements through the auto-rickshaw drivers of Delhi. Roadside hoardings and radio messages were also aplenty. It drew large number of volunteers for campaigning on the streets. "Campaigns on Google Hangouts, Facebook and other social networking sites to woo voters have become very common.

After ruling the state for three consecutive five-year terms, the incumbent INC government headed by Shiela Dikshit was facing a strong anti-incumbency mode. The growing nationwide dissatisfaction on the Central government, also headed by the INC, over growing corruption and rising prices of essential commodities was adding to the anti incumbency burden of the state government. The BJP had already started preparations for the national elections 2014 under the assertive leadership of Narendra Modi and its machinery was well in place to capitalize on the anti-incumbency mood of the voters.

The BJP campaign attempted to gain from the anti-Congress mood banking upon its honest CM candidate for the state, and able prime ministerial candidate as the best alternative for the nation. Through billboards and media messages, the party projected BJP as a better alternative to give reprieve from the unprecedented price hike. The party used online marketing, as websites and social networks are flooded with pro-BJP advertisements.

The INC campaign Claimed that Delhi has progressed by leaps and bounds during its rule. Mass media advertisements, especially television advertisements depicting the developmental work undertaken during its fifteen years rule dominated its campaign. The party attempted to exploit the anti-minority image of the BJP and its prime ministerial candidate while promising through its prime ministerial candidate its commitment to inclusive growth.

Table 3. Delhi Assembly Election Results.

Name of party	Seats contested	Seats projected			Seats won	% of vote share	Vote share Swing
		AAP-Cicero	BJP	Times Now, C-voter			
INC	70	08/14/13	11	24	8	24.55	–15
BJP	66	11/17/13	36	25	31	33.07	–3
AAP	69	38–50	5	18	28	29.49	+24.49

Impact of Social Media Campaign

Accurate quantification of how effective the social media campaign strategies had been in changing voter behaviour is difficult to ascertain.

Optimists claim that "twitter can topple governments" while pessimists suggest that in a country like India it does not really matter.

The Internet and Mobile Association of India (IAMAI) and the Mumbai-based Iris Knowledge Foundation, have reported that the 93 million Facebook users will "wield a tremendous influence" over the results of the polls in 160 of India's 543 constituencies. There could be a vote swing of 3 %–4 % in states, where the internet users are sizeable. The vote swingers can be pre-dominantly young men and non-working women whose affiliation towards social media is high. Perhaps, it is this conviction which has forced the ruling Congress party and the main opposition Bharatiya Janata Party (BJP) earmarking 2–5 % of their election campaign budgets for social media, as claimed in a study by IAMAI and Mumbai-based market researcher IMRB International.

Professor Anand Kumar, sociologist in Jawaharlal Nehru University believes that social media is a new phenomenon, to which men over 50 and women over 30 are not addicted. He explains that only having good foothold on social media doesn't guarantee win in elections, but it impacts 15 to 20 per cent voters in most of the constituencies". Professor Amit Sheth at the Ohio Centre of Excellence in Knowledge-enabled Computing at Wright State University says "with social media data, we can measure sentiments, for example, before a rally, during the rally, and post-analysis.

Aravind Gupta of the BJP claim that in 2014 elections social media affected 30 to 40 percent of the overall seats, a number he said could go up to 60 percent by the 2019 general elections. In 2014 election, in many constituencies, social media was amongst the top three communication tools, overtaking traditional methods such as advertisements. Twitter and Facebook helped break down information barriers between politicians and voters, allowing them to better understand the candidates and the issues they were promoting. Election Commission Officials say social media played an important role in the Delhi assembly elections and ensured the success of Aam Aadmi Party's 'Call Delhi' campaign.

"The social media effect was huge for the BJP… They really understood that social media is an extended version of the campaign trail," said Michael Kugelman, a senior program associate for South and Southeast Asia at the Woodrow Wilson Center. By the time he was sworn in as prime minister, Narendra Modi had more than 16 million "likes" on Facebook, the second most for any politician in the world, and he was the sixth most followed world leader on Twitter. Modi's popularity carried the BJP to victory with 282 seats out of the 543 seats in parliament, the biggest win by any party in India after 1984. Through its use of social media, the BJP was able to sway many young voters.

Sceptical of the impact Pankaj Prasoon, director at the Centre for Indian Political Research and Analysis (CIPRA), claims that "social media is only a political propaganda to woo young voters. Anurag Mittal, National Coordinator at Association for Democratic Reforms (ADR) asserts that "most of the netizens are not voters; they are not even excited about their enrolment as voters." The impact of social media is also subject to the existence of internet infrastructure. In rural areas, the internet penetration being just one-twelfth of urban penetration, as claimed in the McKinsey report, the use of online media is limited. The impacts of online campaign, as of now, remains limited largely to urban areas and well-off sections of society" claims Paranjoy Guha Thakurta, a political analyst.

Sunil Abraham, executive director of the Bangalore-based Centre for Internet and Society, however, is doubtful about the social media's expected effect. Social media "is not a true dipstick. It really is only a dipstick of the elite," he claims. "Illiteracy and limited internet penetration still being a reality in rural swathes of the country, rallies, billboards, and news media advertisements still holds the key to reach the majority of the 814 million-strong electorate".

While admitting that social media played an important role in the BJP's victory, Aravind Gupta acknowledged it only as a part of the overall strategy. "You're not going to win an election only by depending on your social media presence," said Kugelman, from the Woodrow Wilson Center. He pointed to AAP and the Pakistan Tehreek-i-Insaf in Pakistan, political parties that had a strong social media presence but couldn't convert that into votes because of a lack of national presence. "The type of messaging and the target messaging, and the agenda on which the election was fought, won them the election rather than social media alone," said Bharath Gopalaswamy, deputy director of the South Asia Center at the Atlantic Council.

Rediff.com's proprietary data analysis results of the Delhi Assembly elections reveal that dominant presence of voters who are Graduates with high English Literacy, holding white collar jobs and who use the Internet extensively favoured the AAP. However, in constituencies where there are only a few white collar workers, the AAP share declined sharply to 17.62 % but the BJP and Congress gained with the BJP vote share sky rocketing to

56.05%. In graduate concentrated constituencies the AAP share jumped up to 48.74% while the BJP and Congress share plummets. As such, there was a clear indication that in the 2013 Delhi assembly election, Internet savvy voters flocked en masse to the AAP. The BJP had an almost equal effect on Internet savvy and non-Internet savvy voters. The INC, it seems, failed to prevail upon the Internet savvy voters, which may probably be attributed to party's failure to use the social media as compared to its two main competitors.

Decoding public perception on effectiveness of social media

To know the public perception on effectiveness of social media campaign, a survey was conducted in the campuses of three universities. A pre-structured questionnaire was administered among a randomly selected sample of 70 respondents, which comprise 38(54.28%) males and 32(45.7%) females in the age group of 18 to 45. The sample includes 38 respondents in the age group of 18–24, 26 in the age group of 25 to 35 and the rest 06 belonged to the age group 36 to 45. 26 of them are graduates, 36 post graduates and the rest 06 are Ph.Ds. 52 of the respondents are students and 18 are in service. 56 respondents are urbanites and the rest 14 belong to semi-urban localities. The responses obtained through the questionnaire have been presented in tables 4 to 9 below.

Table 4. Where and how they access the Internet.

Place of access	No of respondents	As % of total	Mode of access	No of responders	As % of total
Home	37	52.86	PC	22	31.43
Office/ Institution	43	61.43	Laptop	42	60.00
Internet cafe	02	2.86	Mobile	34	48.57
No where	00	00	Other handhelds	00	00
No response	00	00	No response	00	00

Data in table 4 reveals that 61 percent of respondents access the Internet in offices or academic institutions against 52 percent who do it at home. 60 percent people access the net through laptop followed by over 48 percent who access the net through mobile. Only one-third of respondents access the net through personal computer.

Table 5. Do they own/use the following?

	No of respondents	**As % of total**
e-mail account	44	62.86
Face book profile	55	78.57
Blog	09	12.86
Twitter profile	10	14.28

Interestingly, the number of active Facebook users outnumbers the active e-mail users as is evident from table 5.

Table 6. How many of the following you received during the elections.

e-service	**Nil**	**1–2**	**3–5**	**6–10**	**More than 10**	**No opinion**
Mobile text messages about elections	28	17	19	04	06	00
e-mails about elections	37	11	12	02	04	04
Twits from politicians	47	09	02	02	06	04
Face book postings About elections	24	16	11	08	04	12

From table 6, it is clear that the respondents got more election related messages through Facebook and Mobile SMS compared to e-mail and twits from politicians.

Table 7. Have you done these during the election.

	Yes	No	NA	No opinion
Expressed face book likes	36 (51.42)	24	08	02
Read blogs of political parties	22 (31.42)	26	00	02
Followed twits by politicians	26 (37.14)	42	00	04

Data in table 7 evinces that over 50 percent of the respondents expresses likes of political campaign messages, 26 percent followed twits by politicians and 31 percent read blogs of politicians during the elections.

Table 8. Do social media political communications influence voters?

Sl No.	Response	No of respondents	As % of total respondents
1.	Strongly disagree	04	5.71
2.	Disagree	04	5.71
3.	Nether disagree nor agree	10	14.28
4.	Agree	40	57.14
5.	Strongly agree	10	14.28
	No opinion	02	2.85

As is evident from table 8, an overwhelming 71 percent of the respondents agree that social media campaigns influence voters, however only 14 percent of them strongly agree that social media influence voters during elections.

Table 9. Which of the following influences voters the most?

Sl No.	Response	No of respondents	As % of total respondents
1.	Mobile SMS	14	20
2.	Face book postings	32	45.71
3.	Twits	02	2.85
4.	U-tube	04	5.71
5.	No idea	26	37.14

From table 9, it is clear that according to the respondents, Facebook influences voters more than the other social media elements.

Asked to recall some popular campaign messages they have received during the elections, a host of messages were mentioned. However, among the messages recalled by the maximum number of the respondents are: 'the power of Aam Admi' (22.85%), 'Har hath shakti har haath tarkki' (17.14%), and Abki Baar Modi Sarkar (20%).

Findings and Comments

The advent of social media has not only changed the modes of political communication, but also changed the very nature of political discourse. The power of social media in public mobilization has been widely accepted after the Arab Spring, the Indian Lokpal campaign, and other isolated public uprising in various parts of the globe. However, an exact cause-effect relation between social media political campaign and change in voting behavior is still to be established. Being guided by a number of studies on impact of social media on election outcomes, this study analysed the issue in the context of the Indian general elections 2014 and Delhi assembly elections 2013.

Despite a low internet penetration among the general population, the wider participation of the youth in social media has made it a useful tool in political communication. The high accumulation of young student population with access to the internet and other social media at their institutions make the impacts far more visible in the major cities like Delhi. Further the spread of low cost mobile handsets and other handhelds available in multitude of forms have expanded the social media users. Majority of the student population having been admitted that they have social media accounts, have received political messages through social media, and most importantly having responded to and followed political leaders on social media are adequate enough to believe that social media might have influenced voters during the elections. The respondents also overwhelmingly believe that social media campaign influence voters. Regardless of the debates about its impacts, social media is contributing to further "democratization of politics" in the world's biggest democracy. Beyond the rich and the poor, now the middle class in particular has found politics interesting, and social media has enabled them express their opinion and talk beyond the channels of the mainstream media. As such, the advent of social media has made Scot's proclamation "comment is free but facts are sacred" really meaningful to the common men.

References

Becker, L. B. and Tudor, V., 2011, "Freedom of the Press Around The World", in Beer Arnold S. and John C. Merril (ed.), *Global Journalism- Topical Issues and Media Systems*, New Delhi, PHI Learning Pvt. Ltd, pp. 82–83.

Bowman, S. and Chris, W., 2003, *We Media. How audiences are shaping the future of news and information,* The Media Center at The American Press Institute, retrieved from <www.hypergene.net/wemedia/download/we_media.pdf>, on February 16, 2013.

Bulter, P., 2007, *New Technology, New Voices*, E-Journal USA, Retrieved from <http://usinfo.state.gov/journals/itgic/1207/ijge/butler.htm>, on January 28, 2013.

Butsch, R., 2008, *The Citizen Audience: Crowds, Publics, And Individuals*, New York, Rutledge.

Carlson, M., 2007, "Blogs and Journalistic Authority: the Role of Blogs in US Election Day 2004 Coverage", *Journalism Studies*, vol. 8(2), 264–279.

Cooper, S.D., 2006, *Watching the Watchdog: Bloggers as the Fifth Estate,* Spokane, WA, Marquette Books.

de Beer, A. S. & Merrill, J. C., 2008, *Global Journalism: Topical Issues and Media Systems*, Pearson, p. 504.

Election Commission of India, "Assembly Elections December 2013 Results", Retrieved April 3, 2014.

Election Commission of India, "Assembly Elections December 2013 Results", Retrieved April 3, 2014.

Friedrichsen, M. and Muhl-Benningghaus, W., 2013, *Handbook of social media management*, Berlin, Springer, P.3.

Garza, N., 2006, Blogging Grows as Popular Section in Online Journalism, *International Symposium on Online Journalism*, University of Texas at Austin, retrieved from <http://onlinejournalism.utexas.edu/detail.php?story=148&year2006>, Grosswiler Paul.

Glaser, M., 2010, Citizen Journalism: Widening Worldviews And Extending Democracy, in Allan, S. (Ed.), *The Routledge Companion to News and Journalism*, New York, Routledge, pp. 578–590.

<http://m.ndtv.com/article/india/election-commission-announces-poll-dates-for-five-states-highlights-427827>.

<http://www.ceodelhi.gov.in/WriteReadData/Affidavits/ECICandidateAffidavit.htm> (accessed on November 22, 2013).

<http://www.ndtv.com/elections/article/assembly-polls/delhi-sees-record-voter-turnout-advantage-bjp-say-exit-polls-454562?curl=1386342946>.

IAMAI Report, 2013, "Social Media Campaign Can Swing 3%–4% of Votes", New Delhi, Oct 08, 2013.

Lasica, J.D., 2003, "What is Participatory Journalism?", *Online Journalism Review*, <http://www.ojr.org/ojr/workplace/1060217106.php> (accessed October 10, 2011).

Lowrey, W., 2006, *Mapping the Journalism-Blogging Relationship, Journalism: Theory, Practice, Criticism*, vol. 7(4): 477–500.

One India, 2013, "EC announces election dates for Delhi, MP, Rajasthan, Mizoram, Ch'garh", 4 October 2013, Retrieved 7 October 2013.

Philipose, P., 2013, "Anxieties in The Republic- Media Metamorphosis and PopularProtest", *Economic and Political Weekly*, p. 20.

Pradhan, P. 2013, "New Media Impacts on Journalism: Revisiting the Dynamics of News Production", *The Media Watch*, Vol.4(2), p. 191.

Ray, T., 2006, *Online Journalism- A Basic Text*, New Delhi, Foundation Books.

Rediff.com, "Play with the Sentiment Meter and Interactive Map tools to see how sentiments can change the election outcome", 11 November 2013, Retrieved 11 November 2013.

Rodzvilla, J. (Ed.), 2002, *We've Got Blog: How Weblogs Are Changing Our Culture*, Cambridge, MA, Perseus Publishing.

Schwitty, A., 2011, "New Media Practices In India: Bridging Past and Future, Markets and Development", *International Journal of Communication*, Vol.5, 349–379.

Shafaque, A., 2014, "Noida poll officials bet on FB to attract voters", *The Times of India*, March 02.

Tang, Q., Gu, B. & Whinston, A. B. (2012), "Content Contribution for Revenue Sharing and Reputation in Social Media: A Dynamic Structural Model", *Journal of Management Information Systems,* 29, pp. 41–75.

Vaishnav, A., 2013, "A New Journalist Is Born: Digital Media Has Brought About A Sea Change in The Way News Is Presented And Perceived", *The Times of India*, p. 20.

Witschge, T. and Nygren, G., 2009, "Journalism: A Profession under Pressure?", *Journal of Media Business Studies*, Vol. 6(1), pp. 37–50.

Zhe Deng & Richard Balme

Chinese Micro-Bloggers' Behaviors Towards Twiplomacy: Text Analysis of British PM David Cameron's Micro-blogs

Introduction

Continuous development of New Information and Communication Technologies (NICTs) has brought about significant, even complete changes in E-era (Castells, 2000; Freeman & Soete, 1997), especially for our behaviors, attitudes and relations in political life.

On one side, policy makers and government officials cannot stand aloof from this new trend as they have to conform to the current development or to explore better opportunities. In fact, the governments and the officials do have paid more and more attention to the interactions with the public by using the forms that NICTs users prefer (such as micro-blog, BBS, IM, multi-media applications and so on), so as to disclose the policies and governmental departments information, express their viewpoints, and listen to the appeals from the public, etc. On the whole, both the governments and the officials have noticed the importance to use NICTs into their daily work and the communication with the public. However, do they really have a good knowledge of how to use NICTs and are they using it effectively?

On the other side, the original "black boxes" in the past – such as policy process, corruptions and even officials' properties along with their private life – are opened and shown in front of the public by the ubiquitous NICTs. As a result, through their self-media, the grassroots, such as social media users and micro-bloggers, can know more and more information, understand better and the better about the operations of the government and politics, as well as express their appeals and participate in the discussions on political issues in the virtual public space more conveniently and more actively. So, what are they talking about and how do they express their opinions, requirements and attitudes?

During late November to early December, 2013, British Prime Minister David Cameron visited China and registered an account on sina.weibo.com (the biggest and most popular website which provides service like Twitter in a Chinese version) with more than 500,000 followers, attracting humorous attention and leading to heated discussions from local media (both social media and mass media). The interaction between PM Cameron and Chinese micro-bloggers becomes a good case revealing both Twiplomacy of heads of states and public engagement of NICTs users – the demands side.

So far, most researchers have studied a lot how state leaders build their images on social media or have already tried to explore Twiplomacy strategies for the authorities. However, we think that only user-oriented Twiplomacy could achieve trust, affection and support. Therefore, this paper aims to, from the perspective of NICTs users, describe Chinese micro-bloggers' attitudes and behaviors which are discovered during their engagement in public discussion, expression and interaction, by using the case of Cameron's Twiplomacy in China – empirical analysis of 200 replies on each of Cameron's 12 tweets (2400 pieces of replies in total) during 9 days, since November 29th to December 7th, 2014.

Twiplomacy

On the whole, Twiplomacy focuses on a sort of mutual relations on Twitter.[1] It is also regarded as a new global study of world leaders on Twitter, since Twitter has become "a new way to communicate with world leaders and a way for these leaders to communicate with each other"[2] in the past five years. In this study, we mainly study its function of allowing common users' direct access to world leaders and the mutual interactions between the state leaders and the public.

1 <http://twiplomacy.com/>.
2 idem.

Twitter and Micro-blog

Since being founded in 2006, "Twitter has become a formidable communication tool allowing the broadcast of short messages to millions of followers"[3]. As kind of SNS (Social Network Service), however, Twitter is different from the others (such as Facebook and LinkedIn) because of its core function of information releasing and communication instead of making friends or social networking.

Micro-blog is usually regarded as the Chinese version of Twitter, which was fully operational by sina.weibo.com in 2008. According to the 2012 report by China Internet Network Information Center (CNNIC), there are already 27.4 million micro-bloggers in China in total by June 2012.[4]

It is noteworthy that: Firstly, same as Twitter, micro-blog also allows 140-character bursts of information. However, 140 Chinese letters are generally able to express quite more complete and profound meanings than the same amount of English or other Latin letters of alphabet. That's one of the reasons given by lots of researchers and critics why Chinese SNS users are more actively engaging in public affairs and discussions than netizens from other countries – they are allowed to express more with the same or even less cost of time and effort. The nature of easy operation and efficiency of SNS tools is usually taken account above all.

Secondly, if twitter is international and worldwide, micro-blog is basically used inside China domestically and focuses on issues about China in the most of the time. In other words, the background of Chinese micro-bloggers and the environment in which they are micro-blogging are accordant and simplex in general, compared to twitters.

Government Micro-blog, Official Micro-blog and Leader Micro-blog

Among studies in and about China, both government micro-blogs and officials' micro-blogs have been paid much attention, no matter in communication, public administration, politics or any other disciplines. But leader micro-blog is another thing. However, in some papers, the three concepts "government micro-blogs", "officials' micro-blogs" and "leader

3 <https://twitter.com>.

4 <CNNIC. http://www.cnnic.net.cn/>.

micro-blogs" are always mixed. In this paper, we define them separately and confirm the last one to be the keyword.

Undoubtedly, government micro-blog represents the standpoints of the governments, releasing official information and communicating with the public on behalf the image of the whole department or organization. In general, it merely aims to provide information service to the public.

In contrast, officials' micro-blogs are registered in the name of officials themselves, and often only represents the officials' personal opinions. At the very beginning, only quite a small part of Chinese officials used micro-blogs, so that these micro-blog accounts were easy to attract concentrations from the public. Later on, the administration department demanded all officials' micro-blogs to illustrate that they could only represent their own viewpoints instead of the organizations they worked for. Different from government micro-blogs, officials' blogs not only burst information about the government or policies, but also release more pieces of their comments on particular events, and especially express their affections and show their private life, like dishes for lunch, activities during weekends, gathering with close friends and so on.

So, what is leader micro-blog? We find that leader micro-blog is different from either government micro-blog or officials' micro-blog, but it more or less contains characteristics of the last two – first, leader registers his own micro-blog account in the name of individual but not the whole country; second, usually it's not operated by the leader himself but an intellectual group; third, to some extent, leader micro-blog reflects attitudes of his country and, generally speaking, he represents the image of his government or even his country.

Twiplomacy in China

Twitter or micro-blog is often considered as creative way to increase communication efficiency among all the users as well as important approach to promote propaganda performance for the government and the organizations. Nowadays millions of micro-blog accounts have been registered in China. Not only do Chinese governments and officials use it, but more than 140 foreign politicians have also registered to be micro-bloggers, conducting Twiplomacy by fostering good images in China and reaching out to the Chinese public.

British Prime Minister David Cameron earned more than 632 thousand followers[5] (most are Chinese micro-bloggers) through his micro-blog account on sina.weibo.com and attracted traditional mass media's attention continuously more than a week as well. To some extent, we must say PM Cameron did have made a successful Twiplomacy in China through Chinese version Twitter – Micro-blog.

However, on twiplomacy.com, twiplomacy Asia study even hasn't done any research on China's situation. Therefore, this paper will supply the part of Chinese micro-bloggers attitudes and behaviors towards Twiplomacy under the framework of public energy field theory.

Public Energy Field

Public energy field is a core concept of Fox T.J. and Miller H.T's discourse theory proposed in their book *Postmodern Public Administration: toward Discourse*. They advocate that public issues should be settled in public energy field, which is a place for the common public to state their viewpoints equally, freely and efficiently. In the public energy field, all participants speak, interact and communicate under the following six principles:

a) Honest. The first principle requires all the participants to speak truth instead of telling lies, which is regarded as the base for dialogue.
b) Suitable intentionality. Only focused intentionality is able to ensure that the dialogue and the communication will centre on an appointed event or a main topic.
c) Self-participation. Voluntary participation encourages participants to willingly discuss, argue and act on particular event, in order to solve the problem or reach to an agreement.
d) Effective contribution. In order to make the public engagement effective and efficient, participants should interact and communicate centering on a particular topic, but not wandering off the subject. For example, participants could put forward a creative point of view, know well about some professional skills, or maybe they have gone through relative experiences and so on.

5 Up to June, 2014.

e) Equality. Public energy field provides each group of participants a fair platform to discuss or even argue equally, controvertibly and face to face, especially preventing from monologue.
f) Communication among some people. Among three forms of dialogue – limited people, most people and some people, the first one causes monologue, the second one makes the occasion confused and easily out of order, but only the last form – some people is the best for communication and public engagement, especially for bringing to some solutions and results.

Micro-blog is taken as a public energy field for the public engagement nowadays, (Sun, 2013) in which micro-bloggers could talk freely and equally. However, according to the six principles of public energy field theory, how about Chinese micro-bloggers performances in public engagement indeed? Have they really already reached the standards framed by the six principles? What are the good sides and bad sides?

Data and Methods

To explore the communication and interaction mechanism of public engagement towards Twiplomacy on micro-blog, this paper starts to construct a set of factors based on Fox and Miller's public energy field theory from six orientations mentioned above.

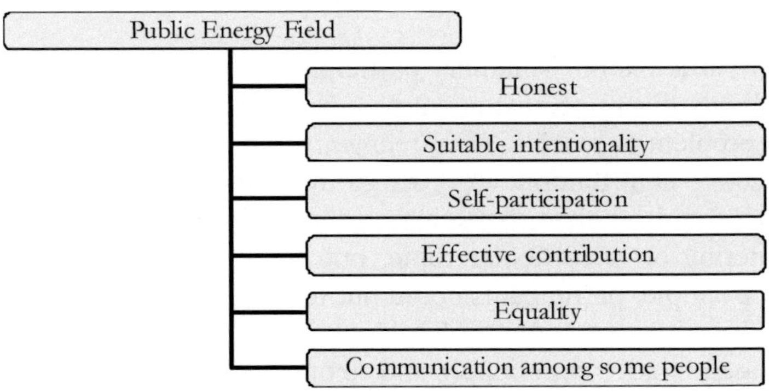

Figure 1: Factors of Public Energy Field.

Through text analysis of 200 pieces of replies on each of Cameron's twelve tweets (2400 pieces of replies in total), this paper works out a description and measurement on Chinese micro-bloggers' attitudes and behaviors in public energy field towards Twiplomacy with data and quantitative analysis.

The twelve tweets are as below, from November 29th, 2013 to December 7th, 2014.

Table 1. Prime Minister Cameron's Micro-blogs.

No	Time	Content	Reply	Retweet	Favorite
1	2013.11.29 17:00	Hello my friends in China. I'm pleased to have joined Weibo and look forward to visiting China very soon. Hello 中国的朋友们，我非常高兴能加入微博。期待不久后的访华！	16361	31438	16450
2	2013.11.29 17:25	I enjoyed visiting the 'Masterpieces of Chinese Paintings' exhibition at the Victoria & Albert Museum in London yesterday before my trip to China. 在访华行程开始前，昨天我有幸在伦敦维多利亚和艾尔伯特博物馆欣赏"中国古代绘画名品特展"。 (5 pictures)	5483	4981	5086
3	2013.11.30 16:09	本周，一组正在学习中文的英国学生来到唐宁街10号做客。他们和我交流了学习中文的心得体会。 [Translation] This week, a group of English students visit 10 Downing Street. They are learning Chinese. We communicate about the attainment experience. (3 pictures)	4044	4091	4112
4	2013.12.2 07:39	已经抵达中国，很高兴这是我作为英国首相的第二次对华访问。稍后我将会和中国的领导人会面。 [Translation] I have already arrived in China. I'm glad that this is my second time to visit China as British Prime Minister. Later I will meet the leaders of China. (1 picture)	5388	7580	7253

No	Time	Content	Reply	Retweet	Favorite
5	2013.12.2 08:45	在现场，我了解到捷豹路虎如何在对中国出口方面获得了巨大的成功。 [Translation] I found on site how Jaguar and Landrover succeed in exporting to China. (1 picture)	1311	953	2059
6	2013.12.2 09:07	我见证了捷豹路虎签署在中国销售45亿英镑的合约。 [Translation] I saw Jaguar and Landrover signing contract of 45 billion pounds selling in China. (1 picture)	3768	4953	3190
7	2013.12.2 20:36	谢谢你们关注我。我很高兴能作为英国首相再次访华。我很想了解你们的想法，所以请留下你们的问题，我会在我的访问结束前回答一些。 [Translation] Thanks for your attention to me. I'm happy to visit China again as Prime Minister. I really hope to know your opinion. So please write down your questions, and I will reply some before the end of this visit. (9 pictures)	21343	42355	13534
8	2013.12.3 22:06	今天再上海一天的行程。 (9 pictures)	8114	5743	5638
9	2013.12.4 18:57	今天下午我在成都访问了龙江路小学。在那里我录制了一段视频，专门回答你们的问题。视频会在接下来的几天发布。请一定要看。感谢你们的提问。 [Translation] This afternoon I visited Longjianglu Primary School of Chengdu. I made a video to reply your questions and it will be posted later. Please watch that. Thanks for your questions. (1 picture)	10720	7796	6879
10	2013.12.5 18:11	昨天，在成都。 [Translation] Yesterday in Chengdu. (9 pictures)	7945	5889	7162

No	Time	Content	Reply	Retweet	Favorite
11	2013.12.6 23:52	对于纳尔逊·曼德拉的逝去，我感到极度悲伤。我的心与他的家人、还有南非及全世界生活因他的勇气而改变了的人同在。 [Translation] I feel very sad about Nelson Mandela's death. I'll be with his family and all people who have been encouraged by him. (2 pictures)	3299	2562	7022
12	2013.12.7 20:14	感谢你们的提问。我在成都时录制了回复。 [Translation] Thanks for your questions. I made a video to reply in Chengdu. (1 video)	13180	33053	14834

Analysis and Discussion

Data analysis reports some interesting results, which reflect the characteristics of Chinese micro-bloggers' public energy field towards Twiplomacy.

Active self-participation

Figure 2 shows high amounts of retweets and replies on Cameron's 12 tweets. Therefore, the conclusion of active participation of Chinese micro-bloggers could be drawn out. In China, social media, especially micro-blog which suits for speedy life and requires no profession criteria, become an important virtual space for Chinese netizens to express their voice and accessing to the authorities. The data below proves the active self-engagement of Chinese netizens in public issues with their self-media – micro-blog.

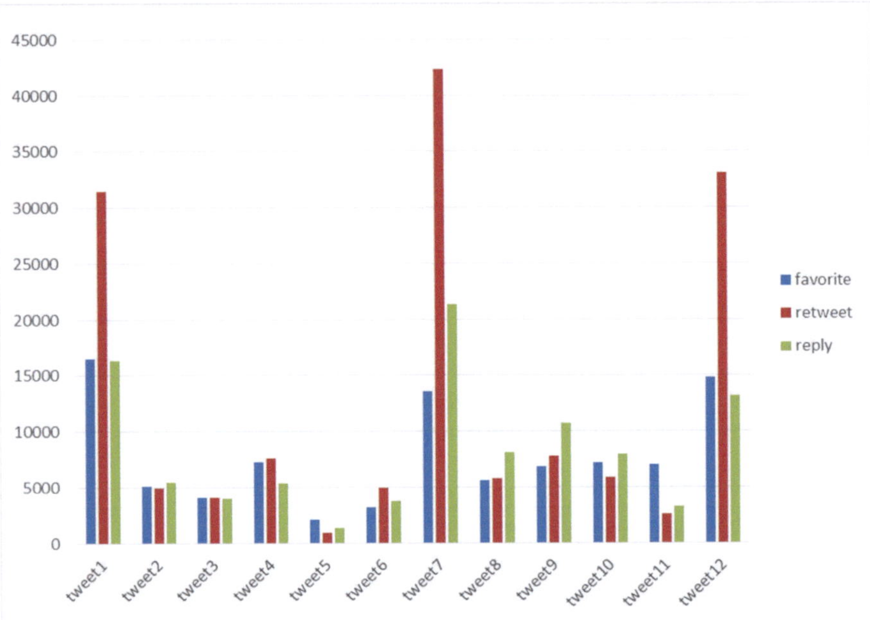

Figure 2: Chinese micro-bloggers' participation in Twiplomacy.

Refers to Twiplomacy, Chinese netizens' initial attitudes towards Cameron's micro-blog account reveals their surprise and curiosity. Micro-blog offers an easy and fresh way for common Chinese people to talk to authorities. That's why more and more world leaders are willing to register accounts on Chinese social media websites and communicate with Chinese – social media do work effectively for communication with Chinese citizens, especially to world leaders, since Chinese micro-bloggers still keep kind of interest and curiosity about the interaction in virtual society with governments, officials and leaders from other countries. It could be said that micro-blog is still a useful tool for Twiplomacy in China.

Lack of effective contribution

However, according to the average numbers of favorite (7768), retweet (12616), and reply (8413), the result shows that there are much more people only gaining information and transmitting information instead of giving information (i.e. expressing opinions and rising demands). (Figure 3)

This phenomenon leads to less contribution and less and less continuation of the conversation between each side, since reading and transmitting original information are lack of new knowledge and standpoints, which causes the communication stops in one side without argument.

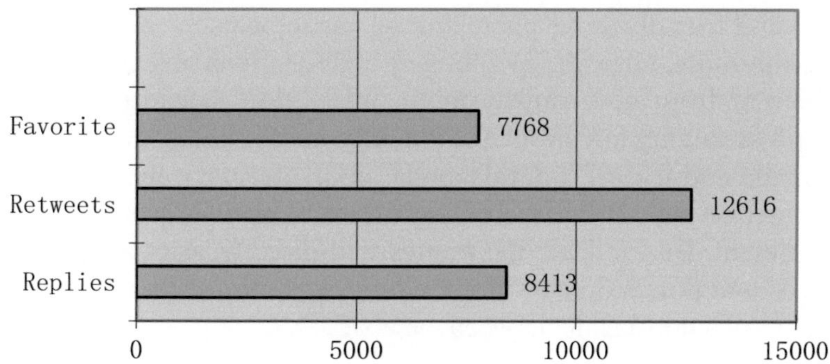

Figure 3: Comparison of Micro-bloggers' Behaviors.

Specialty of Intentionality in Twiplomacy

Partial Suitable Intentionality

Secondly, Chinese micro-bloggers do express their historical and political requirements to foreign leaders in this virtual public energy field, such as calling for Britain's returning Chinese cultural relics back. For the good side, micro-bloggers sometimes can find out several specific topics to start and continue the conversation intentionally. These topics become the subjects of dialogue between the Prime Minister and Chinese micro-bloggers and the aims of their communication and interaction.

For Twiplomacy: Intentionality or Entertainment?

However, for the bad side, most of the topics are irrelevant to the content of Cameron's tweets. In their reply, expression is stronger than dialogue and reply. Some of the irrelevant replies are centered on funny jokes or relaxing topics, aiming for communication and attracting attentions from both the Prime Minister and other micro-bloggers. For example: "Could you teach me English, Prime Minister?" [6], "I want to marry you."[7]

Political issues are much less discussed than the trivial round. However, life related topics online bring about more delighted interactive relationship and friendly atmosphere among participants.

Furthermore, all in all, Twiplomacy and functions of leader micro-blog are different from government-public interaction, because usually they don't aim at solving any problems but only for communication, setting up good image and trust. From this angle, no matter what the topics are, all the interaction replying the bursting micro-blogger is worthy, including entertainment. But, indeed, the replies that directly react and respond to his words and points do contribute more for continuation of the dialogue and reaching a consensus between each side.

Multiple and Personal-Public Intentionality: Honest or Dishonest?

There are also a part of micro-bloggers may disclosure particular public issues which they have no platform offline to speak out. For instance, one micro-blogger accuses the water pollution which kills his family[8], another reports gangland in Wei Fang, Shandong Province[9]. These micro-bloggers are talking about some personal issues which are also problems should be solved by the public administration departments. However, obviously the problems haven't been solved through face-to-face channels.

Among studies on virtual society management in China, disclosure right and disclosure power is always an important component. The public always complain that they don't have chance in reality to suppose their opinions and to participate in policy decision. Emergence of virtual society provides convenient and equal platform for all common people. They finally find a space to protect their right of speaking, but the power of their voice is still a suspicion. Take the two micro-bloggers above as examples. Their replies had been covered by tens of thousand pieces of other replies.

6 <http://weibo.com/u/1985804364>.
7 <http://weibo.com/u/1769816283>.
8 <http://weibo.com/u/3938707057>.
9 <http://weibo.com/u/3938707057>.

It is worth mentioning that, are these irrelevant declarations true or not? Of course not all of the micro-bloggers are honest. Refers to these both personal and public issues, some of the micro-bloggers are indeed telling truth, but there are also some rumors spread just aiming to damage the order of the society and to destroy the image of the government. This is one of the main problems in virtual society management in China.

By censorship and legislation, for most of the time, Chinese government could be sensitive to these falsehoods. On September 5th and September 2nd, 2013, the Supreme People's Court and the Supreme People's Procuratorate discussed and passed the law about libel action on internet, including transmitting falsehoods. This is the first time that the government manages netizens' behaviors on internet with legislation. The most severe ten rumors transmission events happened in the year of 2013 have all been punished and the main participants were sentenced. As a result, the situation has been turning to be better and orderly recently.

Dilemma of "many people": equal, disorder or problem of partial?

Among all the replies, there are lots of pieces of them totally irrelevant to Prime Minister Cameron's tweets or even without any relations to public issues – actually, these replies are commercial advertisements and pornographic information.

Therefore, two points to be noticed: on one hand, there is still a distance between China's micro-blog virtual space and real public energy field, since it is an open platform for the major of the public but not "some people".

Compared to other social network services, micro-blog is easier to operate for most of the netizens. Moreover, the limit of no more than 140 characters requires much fewer deep thoughts or writing skills especially compared to blog and BBS, providing an equal space, to participate in discussions, for all netizens who only need to have an access to the internet and equipments like computers or mobile phones, no matter what their education backgrounds are or how many knowledge or skills they have mastered. Furthermore, diverse forms of text, such as words, pictures, videos, attract netizens to explore it. What is more, micro-blog allows the users to follow and get knowledge of anyone they like and care about without the other side's permission. In other words, the connection way among micro-bloggers is one-way, different from mutual agreement like Facebook. Single connection way provides micro-bloggers a chance to know whatever they want and an approach to whoever they are interested in. In return, micro-blog attracts most SNS users. However, "all people" doesn't always equal to "equality".

Firstly, from the angle of public energy field, quite a large amount of participants is not the best choice for making decision and reaching to an agreement, since it's usually hard to control the situation in order. Along with absence of regulation on social media in China for so many years, it often becomes out of order for public engagement in the virtual society, which is now an urgent problem of public management for Chinese governments.

Secondly, it is obvious that some of the advertisements and pornographic words are produced by computers automatically and in purpose. Therefore, media literacy is another thing need to be discussed. The Chinese word "Suzhi" (quality) is being paid more and more attention in both real society and virtual society studies. When offering a field of public engagement, education, regulation and promotion for better order also cannot be ignored.

On the other hand, another reason why "many people" isn't better than "some people" in Chinese micro-blog space is that, in China, there are still quite a lot of people having no access to micro-blog, such as old generation and people living in the countryside. So "many people" doesn't mean fairness since the voice only represents opinions of parts of the whole public, especially opinions of those enjoying comparatively better living environments.

Conclusion and Suggestion

China's micro-blog provides a basic public energy field for public engagement, referring to its information service to common people and allowance of all micro-bloggers' participation.

However, the difference between China's micro-blog and real public energy field is still obvious from the behaviors of Chinese micro-bloggers: First, Chinese micro-bloggers prefer reading information rather than participating in discussion; second, replies given by Chinese micro-bloggers usually wander off the main subjects rose by the leader micro-blog, which requires more efforts of creation and infection from the world leaders; thirdly, Chinese micro-bloggers do like the form of communication with world leaders on social media, which proves that micro-blog is still a useful tool for Twiplomacy.

Findings of this study and the strategies proposed based on the findings by this paper will help government policy makers or political leaders design and implement policies and strategies to improve online interaction with netizens and propaganda through social media. On one hand, world leaders should take good advantages of China's micro-blog for Twiplomacy, and it's also an inspiration for Chinese authorities – there still hasn't been any Chinese leader appearing on social media. On the other hand, Chinese governments should make great efforts to improve China's online public energy field: a) Establish laws to regulate the order on social media for better public engagement; b) education on Chinese NICTs users for better media literacy in order to create an efficient mechanism of dialogue by the micro-bloggers themselves; c) Guidance to centered subjects is essential and significant nowadays to ensure an efficient discussion and communication in an intentional situation.

References

Castells, Manuel, 2000, *The Rise of the Network Society: Economy, Society and Culture*, Oxford, Blackwell Publishing.

Fox, Charles J., Miller, Hugh T. & Yanhong, Chu, 2002 (ed.), *Postmodern Public Administration: toward Discourse*, Beijing, China Renmin University Press Co.

Freeman, Chris & Soete, Luc, 1997, *The Economics of Industrial Innovation*, London, Continuum.

Zhuohua, Sun, 2013, "Government Micro-blog Corresponds to Public Energy Field", in *Shandong Social Sciences*, pp. 154–157.

Aysun Akan

Making Sense of Gezi Park Protests: Gezi Park Protests as a Defence of Lifestyles

Introduction

The Gezi Park protests turned into a country-wide mass protest against the current Justice and Development Party, JDP, (Adalet ve Kalkınma Partisi, AKP) government, It initially started out as a relatively small protest to prevent the destruction of a park in Istanbul's Taksim Square. The protest quickly mushroomed into a general revolt against the rule of Erdoğan and his ruling JDP. The protests revealed a number of grievances by a wide variety of groups and individuals from different segments of society. One of the outstanding concerns of the protestors was Prime Minister Erdoğan's interference in people's different lifestyles.

In its second term in office, JDP government imposed restrictions on alcohol consumption, especially in public spaces. The Prime Minister Erdoğan in his public speeches frequently insulted alcohol drinkers. Similarly, Erdoğan attended marriage ceremonies; delivering anti-abortion speeches in defence of the proposed new law to restrict abortion rights and making couples promise to have at least three children. The Prime Minister, moreover, began to advocate a more Islamic lifestyle, as the acceptable lifestyle, this created a fear of an Islamisation of society among secular sections of society. A significant number of slogans throughout the Gezi Park protests reflected reactions against the Prime Minister's blatant interference in lifestyles that differed from the conservative Islamic lifestyle which Erdoğan was promoting. In this paper, I will first, attempt to give a brief historical context of Turkish politics and the emergence and the development of Political Islam, which I believe, will help to give us a better understanding of current political developments. The second part will focus on the Gezi Park protests, including a detailed analysis of the satirical cartoons and slogans targeting Mr. Erdoğan's Islamisation policies, followed by the conclusion.

Historical Context

Kemalism, named after Mustafa Kemal Atatürk, the founding father of the modern Turkish Republic (1923), as an ideology based on homogenisation of ethnically and religiously diverse population, aiming, ultimately, to create a modern western state and society. Modernisation of society from top down alienated predominantly peasant and religious society. While radical secularism, one of the main pillars of Kemalist modernisation, has been a strong reference for a significant section of the middle class and civil-military bureaucracy, the majority of Turkish people continue to identify themselves with reference to Islamic culture.

During the National Independence War (1919–1922) and during the consolidation of the nation state in the 1920s and 1930s, the Kemalist state adopted an Islamic discourse to mobilise the masses and secure the alliance of influential religious leaders. However, Mustafa Kemal made a careful "distinction between regressive and progressive Islam, promoting the latter as compatible with modernisation" (Çınar and Duran 2008, p. 27).

Some of the initial reforms carried out in the first two decades after the establishment of the Republic (1923), such as the abolishment of the Caliphate (1924), closing down religious orders and madrasas, based on the teaching of Quran, and introducing the unification of Education (1924), thereby creating a secular education system disempowered the ulema (religious scholars) and ultimately, de-emphasised Islam as part of Turkish republic in order to dissociate the regime from Islam. The main objective of the Kemalist regime was to control the autonomous Islamic political and civil society activities. The establishment of the Presidency of Religious Affairs (Diyanet İşleri Başkanlığı) in 1924 served this objective well as the Kemalist state aimed at "promotion of an acceptable state Islam and the submission of religion to the reason of the state" (Çınar and Duran 2008: 22). In other words, Kemalist secularism differed radically from the Western sense of the word, political Islam was "eliminated from the public political realm", but through the Presidency of Religious Affairs, government aimed at promoting a state-interpreted Islam (Buğra 2002: 189). Thus, until the 1950 general election, the first multiparty election in the history of republic since 1923, the ruling Republican People's Party, RPP (Cumhuriyet Halk Partisi, CHP), imposed a strict control on Islamic political and civil society activities to prevent a threat to the secular republic by, what the Kemalist believed, a backward and reactionary force. The election of the Democrat Party, DP, in 1950, allowed the representation of the 'periphery' in politics, as the DP showed a sensitive attitude towards religious sensibilities of the people.

Nevertheless, despite partial modernisation and transformation of Turkish society in the foundation years and the consolidation of the new republic between the 1920s and the 1940s, as the state's strict control over economy and in social life largely benefited those close to the authorities, the republican mass education system created new social groups who came from the rural areas, through their education could find new paths for social advancement in state bureaucracy (White, 2008). DP allowed these groups to take up high positions in the state bureaucracy (Buğra 2002: 189), which predominantly consisted of secularist elite. Likewise in politics, the DP provided a refuge for these new groups with Islamic sensibilities. These new social groups distinguished themselves by "their values and lifestyles informed by Islam as a frame of reference for daily life" (White, 2008: 350). The 1960 military coup ended the DP period, however, the Islamists got a foothold in the Justice Party, JP, (Adalet Partisi, AP), governing party between 1965–1971 and the major partner in coalition governments throughout the 1970s, claimed to be the follower of the DP.

It was, however, in 1970 that political Islam was fully articulated by the newly founded National Order Party, NOP, (Milli Nizam Partisi, MNP). The party represented the economic interests and values of the petty bourgeoisie of Anatolia and, then, the strong professional and urban mercantile middle class with roots in small towns and villages who found an opportunity for political advancement. The party programme included modernist economist development, promising to construct dams, build missiles and so on, in the name of the glorious Ottoman past and God. Thus, despite the NOP's open Islamic agenda, advocating transformation of Turkish society in line with Islamic principles, "this class antagonism was expressed in the cultural idiom of cultural differences" (Çakır 1995: 21; White: 191–2).

The NOP was banned in 1971 by the Constitutional Court, but Erbakan, the founding leader of the National Outlook Movement, NOM (Milli Görüş Hareketi) in 1969 and the first influential leader of Political Islam in Turkey, founded National Salvation Party, NSP, (Milli Selamet Partisi, MSP) in 1972. The party's success in the 1973 general election made it a coalition partner with the RPP. The main motto of the NSP was "morality and ethics followed by developing heavy industry" (Çakır 1994: 23). The 1980 military coup banned all political parties including the NSP.

The 1970s were characterised by the organised working class actions and the flourishing of left-wing political organisations, including a strong student movement. The period also witnessed the violent clashes between the left and right wing groups, which provided a pretext for the 1980 military coup. The 1980 coup eradicated the left opposition and the organised working class action; banned unions prosecuted the left-wing activists. Although all forms of political expression were banned, the ultimate target of the coup leaders was the left opposition. 'The leaders of the coup adopted "Turkish-Islamic synthesis" as their ideological position to consolidate conservative forces in order to pacify social dissent' (Buğra 2002: 189). Thus, Islam and the Islamic identity became a significant feature of the political and social life in the 1980s Turkey (Çakır, *ibid*).

In 1983 Welfare Party, WP, (Refah Partisi, RP) was founded after the junta lifted the ban on political parties. However, the great breakthrough for WP came in the 1990s; the WP turned into a mass party with over one million members in 1993. In addition to the fertile ground provided by the military coup's ideological restructuring of political and social life based on Islamic and conservative values, further enhanced by the Özal years. Turgut Özal was Prime Minister between 1983–1989 and President between 1989–1993 of Turkey. Özal formed the Motherland Party (Anavatan Partisi, ANAP) in 1983. Motherland Party became a melting pot of centre-right, centre-left, the ultra-nationalist and the Islamist views (Heper 2013: 143–44). Although Özal distanced himself from the secularists and radical Islam; he publicly expressed his Muslim identity, attending Naksibendi Brotherhood's İskenderpaşa Dergahı (Seminary) in Istanbul. Although the religious orders were outlawed during the single party era (1923–1946), the Nakşibendi and the Kadiri orders remained active on the Islamic scene. The late Prime Minister and President Turgut Özal was close to these circles and promoted businessmen of Anatolian origin who had close ties with these orders.

In addition to the above favourable conditions, the success of the WP was partly the result of a well established recruitment system, an organisational ability and a new strategy of separation of religious matters from its party activities. The focus on winning the elections and expanding the membership base required the party to prioritise economic and social programmes (Çakır 1994: 58). This is not to say that the WP abandoned religious demands and vocabulary, on the contrary, it adopted a confrontational style in its dealing with the secular establishment, like the bureaucracy and the military. However, the WP also advocated an Islamic life style, as defined by the WP, for all nominal Muslims, disregarding their preferences and ignoring the fact that practicing Islam had a number of variations in Turkey, quite different than what the WP preached to. As Cizre (2008: 1–10) argues, the WP targeted the idea of secularism as much as the secular establishment, in order to mobilise Islamic values. The WP was closed down in 1997, as a result of military pressure, what is known as the 28 February process for violating the secular principles of the republic.

The 1990s represent a significant turning point both for Turkish politics and the Political Islam. It was in the 1990s that Political Islam became more visible in political and social spheres; "the success of the WP in 1994 municipal (gaining 19 percent of the vote) and in 1995 general elections (with 21.4 percent of the vote nationwide), the formation of Islamic foundations, an Islamic labour-union confederation and an Islamist businessmen's association, the proliferation of pro-Islamist TV channels, newspapers" (Gülalp 2001: 433) and a series of demonstrations by the female university students demanding the right to wear Islamic headscarf's on the university campus grounds. The veiling issue entered the political agenda of both Islamists as the most pronounced cultural expression of the new Islamist political subjects (Yavuz 2003: 208) and to many secularists this was seen as an alarming sign of the increasing strength of what they saw as reactionary Islam. Hence, the issue of identity became a political battleground between the secularists and the Islamists.

Politicisation of Islamic women was partly the result of the increasing presence of the new Islamic middle class in the party, in which Islamist women played a significant role. Involvement in protests against the headscarf ban in universities and party political activism politicised Islamist women who, especially in the 1990s, 'began to create new public spaces confirming to the basic Islamic requirements, such as Islamic hotels, cafes, hair dressers and so on (Çayır 2008: 69).

JDP, in addition to the traditional support base of the Islamists, the small-scale commercial and industrial bourgeoisie, appealed to the second generation bourgeoisie, emerged as a result of the globalisation of the economy and the structural changes in the Turkish economy in line with the globalisation trend. As the "global trend allowed forms of 'flexible accumulation', include a rise in a smaller-scale manufacturing and self-employment" (Gulalp 2001: 436 in Bozkurt 2013: 381). This newly arising Anatolian bourgeoisie became a significant force behind JDP aspiring to compete in European and the world market (Çınar and Duran 2008: 32). As Bozkurt (2013: 381) explains from the 1990s onwards Islamic capital has taken advantage of the export orientation of the economy with the result of emerging of some holding companies as powerful as the "core" capital in Turkey (Gulalp 2001: 444 in Bozkurt 2013: 381). Thus, rather than the traditional discourse of the Political Islam of the 1970s that emphasised social justice, redistribution and heavy state involvement in economy, the JDP, supportive of free market principles, appealed to the second generation bourgeoisie (Bozkurt 2013: 381). Like the traditional support base of the Political Islam consisting of small scale commercial bourgeoisie and the new middle classes this "second generation bourgeoisie" also "reframed their religious demands and ideals in terms of human rights and liberties" (Çınar and Duran 2008: 32) and identified themselves in terms of terms of Islamic culture and conservative life style.

JDP's reference to Islamic values and the conservative life style also appealed to the housewives, farmers, blue collar workers and the unemployed, of which the JDP gained the majority of its vote (Bozkurt 2013: 374). However, populist policies seem to be as significant as the ideological appeal of the JDP for these groups. JDP's "social assistance programmes, which were increasingly used as substitutes for welfare state functions" aimed at relieving the poor. Social assistance programmes, funded by extra-budgetary funds, free from parliamentary approval and gives a considerable power to the central government in allocating revenues (Bozkurt 2013: 374). However, JDP's commitment to the free market economy, consisting of a tight fiscal policy, consistent with the macroeconomic stabilization agenda of the International Monetary Fund (IMF), it was no surprise that the JDP arose triumphantly after the split of the WP into the traditionalist Felicity Party (Saadet Partisi, SP) and the reformist JDP in 2001. The landslide victory of the JDP in the 2002 general election, winning two thirds of the seats in the parliament, opened a new period in Turkish politics. The Prime Minister Erdoğan said that he took of his "National Outlook shirt", by which he meant that he abandoned the Islamic political project, and defined his party's line as "conservative democracy" (Çayır 2008: 62). JDP abandoned Islamic vocabulary and underlined human rights, social justice and economic development, enjoying the support from a wide section of the population representing people from all strata in its first term in office. The JDP managed to embrace new urban classes, who were excluded from cultural and economic life by the Kemalist elite, the rising Anatolian bourgeoisie seeking business opportunity at the centre and poor urban and rural population who found a JDP's commitment on social justice and emphasis on Islamic identity attractive, and finally a significant section of the Kurdish people who put his faith in the JDP initiated peace process which was the first civil attempt to end the war between the PKK and the Turkish military and to recognise Kurdish ethnic identity. However, after the 2007 election victory the JDP, after passing a series of legislations to eliminate the role of the military in politics, adopted a more Islamic vocabulary. This was accompanied by a gradual Islamisation through de-secularisation of the state and society; the lifting of the headscarf ban in public service, reforming the national curriculum to allow more religious instruction in order to eradicate Kemalist mythology, restrictions on abortion, banning of alcohol consumption in open public spaces, restrictions on sales of alcohol in certain districts, reducing the primary education to four years instead of eight, prohibiting chanting of the Kemalist oath at schools and ending the official celebrations of National days, such as the beginning of the National Independence War, the opening of the Grand National Assembly and the commemoration of Mustafa Kemal Atatürk's death.

In addition, the JDP restructured the TOKİ (Housing Development Administration of Turkey), which undertook massive construction project under the name of urban renewal and provision of housing for the poor; construction licences were granted to the construction companies which were owned by the relatives or close associates either of the Prime Minister or other members of the government. The urban renewal projects opened valuable and highly profitable lands at the heart of the cities for the pro-government constructors, while the previous habitants were resettled to tower blocks built at the edges of various cities. One of the government's other major projects was to build over two thousand dams. The majority of these were planned to be constructed in the Black Sea region, which is considered to be one of the world's natural reserves, this drew condemnation from the local population and many environmentalists.

The JDP government, thus, alienated half of the population; the other half is considered by the Prime Minister to be his loyal supporters referring to his forty-seven per cent electorate base. The government's plans for Taksim square seem to serve, first, the Islamisation of public space through changing the Republican symbols that dominate the square, like the National Independence monument at the centre of the square and the Atatürk Cultural centre overlooking the square, to Islamic symbols, like a mosque and an Ottoman era barracks. Secondly, to grant lucrative construction contracts to his close allies in the most important commercial centre of Istanbul, and finally, like all the right wing governments in Turkey, the JDP, wanted to control the square, a place which have been a gathering place for all political protests including massive May Day rallies and various other protests and demonstrations by different political groups and civil society institutions. Thus, the JDP awoke the deep-seated fear in the secular minded-people by its Islamisation polices; created an anxiety in liberal minded-business circles by creating a clienteles business culture and caused discontent among environmentalist by opening natural reserves to business.

Gezi Park Protests

On the 28th of May protests began in Gezi Park in Taksim Square at the heart of Istanbul when the government officials began to destroy the park to redevelop the square. They planned to build an Ottoman-era barracks, a shopping mall and a mosque on the site. The protest turned into a mass unrest on 30th and 31st of May when the police raided the park to clear the protestors who were occupying the park. The violent clearing of the park protesters triggered country-wide demonstrations and protests which went on throughout June 2013.

The Prime Minister Erdoğan showed no tolerance towards the protestors and in a speech to the parliament, blamed the protestors for being pawns of an international conspiracy against Turkey. He went further and described the protesters as terrorists and vandals, calling the protestors as "çapulcu", riffraff. The protestors adopted the word "çapulcu" and many people changed their face book names or surnames to "çapulcu".

The absolute silence by the mainstream media in the week of the greatest unrest Turkey experienced in decades was due to direct government interference in media, repressive press laws weak trade unions, self-censorship of the journalists under pressure of media owners who has close business ties with the government (Tunç 2013: 12–13). In the absence of Turkish mainstream media coverage people turned to foreign and social media. Erdoğan attacked the foreign media for manipulating the news and damaging the reputation of Turkey, thus serving the conspirators. It was under these conditions that the social media provided a platform where people could express their opinions about the protests and get information. In other words, citizen journalism grew. The participants at the protests used live stream channels such as Ustream and live-chat-tools, quickly updated information as well as generating content, thus fulfilling the gap left by the traditional media's censor. Despite misinformation and the lack of verification, hate speech, social media provided a powerful means for the participants to mobilise action and raise awareness (Tunç 2013: 12–13).

A significant number of slogans, messages, photos and cartoons sent through face book, twitter, video share and so on targeted Erdoğan's conservative views and policies through irony, satire and mockery, which infuriated Erodoğan, who listed twitter, after terrorism as the biggest menace to society. "The number of tweets sent daily increased by 370 percent, from 7.2 million to 9.6 million during this period" (Tunç 2013: 12–13). According to a Guardian report (18 June 2013), two millions tweets were sent on the first day of the protests, on 28 May between 4pm and midnight.

The protestors represented a wide variety of classes and social groups: football fans, environmentalists, workers, students, women, LGBT (lesbian, gay, bi-sexual and transsexual) groups, Kemalists, Alevis (the largest non-Sunni group in Turkey), Kurds, artists, leftists, and trade unions. It was a unique moment in the history of Turkey that such a diverse and politically opposite camps came together on a protest against the government policies. JDP government has been facing criticisms by environmentalist who are concerned with government's disregard of the devastating environmental costs of Hydro-electrical power plants in Turkey's natural reserves; women who are frustrated by government's interference in their lives such as restrictions on abortion; gays, lesbians and transsexuals who face legal discrimination and social pressure; urban poor who have been relocated to the outskirts of Istanbul as a result of government's urban renewal programme; Kemalists who see government's interference with people's lifestyle, such as restrictions on alcohol consumption and banning alcohol consumption in open public spaces as signs of Islamisation of society; workers who are concerned with the curbing of the bargaining power of the workers at the hands of the pro-government unions; students whose demand for a more democratic university suppressed by government. Despite different agendas and concerns motivated the protestors, they all shared a common concern, the increasing authoritarian policies of the government and Erdoğan's unchallenged authority.

Cartoons and slogans at Gezi: A Battle over conflicting identity claims

Although Gezi Park protests cannot be reduced to 'a battle of identities', a significant number of cartoons, slogans, pictures and messages demonstrate sensitivities of the people who strongly identify themselves with the Western values and indicate a concern over Erdoğan's conservative policies. For example a cartoon, depicting Erdoğan in Islamic headwear with a dollar sign on it, illustrates well the common criticism against government's capitalist policies sugar coated with Islamic vocabulary. Similarly, another cartoon which was widely circulated during the protests, showed Erdoğan in a traditional Arabic long white dress and a headwear standing on a desert, saying "once we get rid of these few trees left over here, the place will look just like Saudi Arabia". This cartoon brings back all repertoires of Republican view of Arabic countries as uncivilised and backward, and demonstrates the secularists' fear that the Islamists will turn Turkey into an Arabic country.

Another cartoon with a caption "four pillars of Erdoğan's policies", four pillars depicted as four fingers pointing at Erdoğan which reads:

"Violent neoliberalization of the Economy", "Blatant Putinization of the Regime", "Hardcore Gentrification of Urban Space" and "Forceful Yozgatization of Social and Private Life". The cartoon criticises economic policies of the government including a criticism of the urban renewal projects and the last finger, or the last pillar of the Erdoğan's regime of forceful Yozgatization of social and private life, points to Islamisation policies of the JDP, as Yozgat, is a religious and conservative town in inner Anatolia, as such it implies that Erdoğan want to turn whole Turkey into Yozgat.

Many secularists criticise Erdoğan's authoritarianism and his attempts to tailor the fabric of Turkish secular society in accordance with Islamic rules; restrictions on the sale and consumption of alcohol in open public spaces are seen as signs of Erdoğan's forced Islamisation of society and his intolerance towards diverse and alternative lifestyles. A slogan written on a pavement, reading "we kiss round the clock Tayyip", is a response to restrictions on kissing in public, enforced by governors, especially in small university towns, where the university students are accused behaving indecently and disregarding moral codes and Islamic sensitivities of local people.

Another slogan reads; "Diren RAKI, dayan BALIK, örgütlen AYVALIK", which translates as "resist RAKI (Turkish traditional drink), hang on FISH, organise AYVALIK" (a seaside town and a popular holiday site). This slogan is a response to restriction on alcohol consumption in public places and alcohol sale. Drinking alcohol is prohibited in Islam, and the Prime Minister often complained about the level of alcohol consumption in Turkey, increased the tax on alcoholic drinks. Moreover, in many occasions, he insulted drinkers, especially of the raki drinkers. Raki is a traditional Turkish alcoholic drink. Drinking raki is ceremonial in Turkish society, raki table is set with lavish food, fish being an essential part of it. Raki table gathering is a kind of entertainment that is closely associated with secular people.

A handwritten message written on a piece paper hanging on a tree read "It is quite normal to regard parks as whorehouses for the stupid who are turned on by watching an udder in a milk commercial". The slogan represents Kemalists' perception of the Islamists as sexually repressed and has an obsession with sex and thus slandering the secularists for having modern relationships as in the West. This view is expressed in a twit message which said "Taksim has been more crowded than a New Year celebration and not even one woman was sexually harassed. I think you should keep some people at home". Taksim new year celebrations usually hit the headlines with sexual harassment incidents, for which both the public and the media blame young men with Islamic and conservative upbringing from the outskirts of the city, who take advantage of the crowd and behave perversely. The twit message , while referring to the infamous Taksim new year celebrations also responds to Prime Minister Erdoğan's claim that he hardly keeps fifty percent at home, meaning fifty percent that voted for him as his loyal supporters, who, at his call would come out and confront the protestors at Taksim Square. Thus the message equates the supporters of Erdoğan with the perverts come out to Taksim at new year celebration and harass women.

Conclusion

JDP's electorate success in 2002 and 2007 general elections and its popularity raised the hopes of a more democratic politics and society in Turkey. This optimism related to two observations; first, the subordination of military to civil authorities, through reducing the role of the Army in politics, and the second, once in power Political Islam becomes moderate. In fact, in its first term in office, then the Prime Minister, Erdoğan, explicitly advocated more freedom of expression for all, social justice for the disadvantaged and more democracy in line with the European standards. However, in his second term in office, Erdoğan has tended towards religiously based social engineering. His plans to restrict abortion, to criminalise adultery, banning alcohol consumption in public places and so on created a concern, if not a deep fear, among the secular minded people in particular that Erdoğan was paving the way for Islamisation of society. Among the millions of slogans, statements, comments and messages, the issue of Islamisation of society and the conflicting identities, secular vs Islamists, seemed figure a significant part during the Gezi Park protests.

The Gezi Park protests, in many respects, were novel. Firstly, the 1990s generation first time involvement in protests; secondly, the protesters were overwhelmingly from educated, western, middle class sections of the society; and finally, humour dominated the protests. One of the main target of humorous slogans was the government's intolerance toward diverse lifestyles. The satirical images of Erdoğan and Islamism posed the age old question in Turkey: does Turkey want to be a modern western country or turn to an Arabic country? The answer is provided in the cartoons by depicting Islam as non-modern and the Arabic countries as uncivilised. Satirical images targeted Mr. Erdoğan not only as a dictator but also as a reminder the past order, the Ottomans, which the modern Republic and Kemalists made every effort to remove from the memories of people.

References

Aydın, Ertan, and Dalmıs, Ibrahim, 2008, "The social bases of the Justice and Development Party", in Cizre, Umit, (ed.), *Secular and Islamic Politics in Turkey, The Making of Justice and Development Party*, London, Routledge, pp. 201–222.

Bozkurt, Umut, 2013, "Neoliberalism with a Human Face: Making Sense of the Justice and Development Party's Neoliberal Populism in Turkey", in *Science & Society*, Vol. 77, n° 3, pp. 372–396.

Cizre, Umit, 2008, "The Justice and Development Party: Making choices, revisions and reversals interactively", in Cizre Umit (ed.), *Secular and Islamic Politics in Turkey, The Making of Justice and Development Party*, London, Routledge, pp. 1–14.

Çakır, Rusen, 1994, *Ne Şeriat Ne Demokrasi,* Istanbul, Metis Publications.

Çayır, Kenan, 2008, "The emergence of Turkey's contemporary 'Muslim Democrats", in Cizre Umit (ed.), *Secular and Islamic Politics in Turkey, The Making of Justice and Development Party*, London, Routledge, pp. 62–79.

Cınar, Menderes and Duran, Burhanettin, 2008, "The specific evolution of contemporary political Islam in Turkey and its "difference" ", in Cizre Umit (ed.), *Secular and Islamic Politics in Turkey, The Making of Justice and Development Party*, London, Routledge, pp. 17–40.

Fradkin, Hillel and Libby, Lewis, 2013, "Erdoğan's Grand Vision Rise and Decline". Retrieved from <http://www.worldaffairsjournal.org/article/Erdoğan's-grand-vision-rise-and-decline>.

Heper, Metin, 2013, “Islam, Conservatism, and Democracy in Turkey: Comparing Turgut Özal and Recep Tayyip Erdoğan”, in *Insight Turkey*, Vol. 15, n° 2, pp. 141–156. Retrieved from <http://file.insightturkey.com/Files/Pdf/15_2_2013_heper.pdf>.

Matthews, Owen, 2012, “In Turkey, Abortion Law Sparks New Battle Over Mosque Versus State”. Retrieved from <http://www.thedailybeast.com/articles/2012/06/06/in-turkey-abortion-law-sparks-new-battle-over-mosque-versus-state.html>.

Tunç, Asli, 2013, “Can Pomegranates Replace Penguins? Social Media and the Rise of Citizen Journalism in Turkey”. Retrieved from <http://freedomhouse.org/sites/default/files/Can%20Pomegranates%20Replace%20Penguins%20-%20Social%20Media%20and%20the%20Rise%20of%20Citizen%20Journalism%20in%20Turkey.pdf>.

White, Jenny, 2008, “Islam and Politics in Contemporary Turkey”, in Kasaba, Reşat, (ed.), Cambridge History of Modern Turkey, Volume IV, Cambridge University Press, pp. 357–387.

Yavuz, M. Hakan. 2003, *Islamic Political Identity in Turkey*, Oxford, Oxford University Press.

Abhay vir Singh Kanwar

Theorizing Cyber activism: Drawing from Durkheimiam lenses[1]

To be here, is to be seen, to see and to see oneself in the act of being seen
Nalika Gajaweera (2006)

Gone are the days when the peer group of the children consisted of neighborhood children. Gone are the days when the sources of recreation were outdoor gains. Gone are the days when books were the only source of knowledge. Today is an era when you are click away from the person sitting across the seven seas, an era when every thought idea can turn into reality with a flick of an eye, an era when extremes meet and represent themselves in a collective conscience to manifest the air of social solidarity. None but phenomenon of cyber activism had made it possible by implementing internet as a cardinal concept.

Reciprocating to above scenario, the undertaken academic pursuit attempts to theorize cyber activism which in recent years has become one of the most fundamental designs of communication. The paper will attempt to delve into intricacies of cyber space and would expostulate how it is not only mapping, gauging and molding the various strata's of society but also cutting across different type of diversity of people across globe. In a academic demeanor, sociology is the one discipline which takes note of the social aspect of society. Tracing the roots of sociology, Durkheim is one of the foundational thinkers who have rigorously theorized the patterns

1 Mapping the genesis of this submission, my heartiest thanks goes to Prof. Susan Vishvanathan, for she not only steered this submission but also encouraged to pursue this academic probe during my stay in JNU Delhi followed by Prof Karunamay Subuddhi in IIT Bombay who offered an intellectually stimulating course of state polity and society. Apart from that I am more than thankful to Prof. Sybi K. George, my philosophy guide under whom I did my R&D for making it the way this academic probe look likes. Including my friends, I would like to extend my heartfelt thanks to Anand Jose Manarkattu, Niraj Nirola, Ubaid Mushtak and Parimal Chahandeqrious who are holding me high in all my academic crust and gruff.

of human socialization in society. The leading assertion of the scholastic Endeavour would be to map the location of cyber space in today's world and scrutinize how Durkheim's initial orientation to study the society in terms of substance and method which he did two centuries back can investigate the coeval reality which manifest dynamic patterns of existence fanned by information grid. Stretching it further the paper will engage, how cyber space reproduces the attributes of "social fact", one of the eminent conceptions of Durkheim. Collective conscience which ensues into social integration which establishes primacy of society over individual's is also one of his prepositions. As cyber world had been a witness to unprecedented swell in last decade, impact of cyber grid on society's collective conscience will also be the focus of study.

The concepts like division of labor and social differentiation are proposed by Durkheim. On the same hand bent of internet with reference to market orientation has triggered millions of diverse job opportunities, taking note of cyber crime in the form of pornography, piracy and digital divide etc one is reminded of the concept of anomy, which in Durkheimiam terminology is the chronic state of affairs in the modern socioeconomic system. Suicide is one of the extreme forms of anomy, how internet helps an individual in connecting to the other individuals and how internet makes an individual alone or too much connected which results into suicide will be delved into. On the wrapping note Durkheim's fascination to engage with the causality of method to investigate the social phenomenon will facilitate us in discerning the reality of cyber phenomenon in this intellectual probe.

Evolution of internet

It was way back when the history of the Internet started in the 1950s and 1960s with the development of computers. This began with point-to-point communication between mainframe computers and terminals, expanded to point-to-point connections between computers and then early research into packet switching. Packet switched networks such as ARPANET, Mark I at NPL in the UK, CYCLADES, Merit Network, Tymnet, and Telnet were developed in the late 1960s and early 1970s using a variety of protocols.

The ARPANET in particular led to the development of protocols for internetworking, where multiple separate networks could be joined together into a network of networks (Romualdo Pastor-Satorras and Alessandro Vespignani 2004. P. 21). In 1982 the Internet Protocol Suite (TCP/IP) was standardized and the concept of a world-wide network of fully interconnected TCP/IP networks called the Internet was introduced. Access to the ARPANET was expanded in 1981 when the National Science Foundation (NSF) developed the Computer Science Network (CSNET) and again in 1986 when NSFNET provided access to supercomputer sites in the United States from research and education organizations. Commercial internet service providers (ISPs) began to emerge in the late 1980s and 1990s. The ARPANET was decommissioned in 1990.

Growth of internet

The Internet was commercialized in 1995 when NSFNET was decommissioned, removing the last restrictions on the use of the Internet to carry commercial traffic (Romualdo Pastor-Satorras and Alessandro Vespignani 2004: 23). Since the mid-1990s the Internet has a drastic impact on culture and commerce, including the rise of near instant communication by electronic mail, instant messaging, Voice over Internet Protocol (VoIP) "phone calls", two-way interactive video calls, and the World Wide Web with its discussion forums, blogs, social networking, and online shopping sites. It is estimated that in 1993 the Internet carried only 1 % of the information flowing through two-way telecommunication, by 2000 this figure had grown to 51 %, and by 2007 more than 97 % of all telecommunicated information was carried over the Internet. In this way the internet grew into a new emerging contemporary reality. Scenario of such sort has entailed into 21st century in which its first decade had been a witness to a robust growth of internet. It has revamped the whole social structure of the society and has fabricated each and every remote and secluded corner of the world by diluting all the economic social and political boundaries. Giving a way to the phenomenon of cyber activism. From early teens to the oldies of late sixties, from rural pockets to the people of the cosmos internet has wrapped everybody in its fold giving a firm base to the notion of collective conscience and social solidarity.

As mentioned above that cyber world is adding to the socioeconomic and political vocabulary of the society, it is completely changing the generic orientation about relationships amongst the individuals. In the academic demeanor, sociology is the one scholastic endeavor that has mirrored the structures of human society at macro level. Tracing the roots of sociology, one can discern that Durkheim has rigorously theorized the patterns of socialization amongst the members of society. As in the section preceding, the growth and evolution of internet has been dealt in, the proceeding probe ahead will assert some elemental conceptions posed by him and how it is replicated by the attributes of cyberspace.

On a Durkheiman note, there are two leading assertions; the first is primacy of social over individual which is followed by the comprehension of society with reference to the scientific temperament. With reference to the two above mentioned conceptual postulates, to Durkheim, social fact, in the discipline of sociology is one of the most fundamental conceptions which with the help of objectivity in the discipline can be treated as things to comprehend the social reality. These social realities have attributes apart from their conceptual apparatus which can only be deduced with the help of empirical investigation rather than a priori reasoning and intuition. It exists independent of the individual's will and is immune to social structure. It characterizes as an object which is definite and leaves no room for subjective interpretations, personal observations (Durkheim Emile, 1982: 3). The forms of social reality implicit in social facts, to Durkheim inhere into material and nonmaterial forms. Where material forms are directly observable E.g. structures of architecture, tangible forms of technology etc. When these material forms manifest larger and more powerful realm of moral forces that are at least equally external to individuals and coercive over them. These are nonmaterial social facts (Ritzer George, 2011: 79).

The above mentioned fundamental attributes of social fact can be set as a marker to adjudicate the intricacies of cyberspace. As in the preceding centuries the individuality was subsumed within the dialectics of state and nation. With the advent of 21st century, the inherited institutions of modernity and industrial society are diluting at an unprecedented pace and distinctiveness is more or less controlled and regulated by the global networks fanned by cyber grid. Everything that comes under time and space is confronting the socioeconomic structure entailed by this phenomenon. Rather it has further supplanted the conventional orientation with reference to geography, social, economic and technical. It has fostered singularity of interactive cultures and has subsumed the globe in its intact. As cyber space is primarily governed by large-scale commercial service providers, they sell its access in a way to captivate the audience in the commercial maze. Hegemonic nature of the cyber content additionally unseat an individual to merely a customer, contriving his personal preferences and manufacture the societal collective conscience which ensues into consumer carnivals.

As social fact is a thing which facilitates in comprehending society with a scientific temperament, correspondingly virtual hole of internet also manifest the same demeanor when cyber activism in the form of Awareness/advocacy, organization/mobilization, and action/reaction raise the concern for any pressing issue of societal apprehension by acting as a transparent medium which is preceded by global deliberations. An apt example could be of social networking sites. They not only help in connecting individuals from diverse milieu but also platforms imperative social issues which expedites their understanding and endow it to a global community. The role of cyber space as a social fact becomes self-evident in various revolutions of economic as well as political nature. In the Arab spring as well as in the campaign "India against corruption", internet played a cardinal role in spreading awareness amongst people. It further resulted into understanding of the political scenario by the masses. Collective conscience beyond individualism was manifested when awareness entailed into mobilization and organizational resistance to existing structures. Analogous to the social fact, internet assisted in deducing the social reality, external to the individual and it was reciprocated by action and reaction by various stakeholders involved.

On a functional note, the information on the internet is divided into data packets and sent to special devices called routers. It transfers the packets across the Internet's different networks. Such networks of communication system are recognized by a complex interplay of communication protocols, hardware infrastructures, and connectivity is ensured by several softwares (Romualdo Pastor-Satorras and Alessandro Vespignani 2004: 1). The operations of internet mediate amongst various tangible as well as intangible structures. Gadgets which help in accessing the internet encompass the tangible domain on the other hand the virtual space encompassing data packets carrying information with the help of electromagnetic waves comprise of intangible realm. Both tangible as well as intangible spheres of the cyber world as similar to material and nonmaterial social fact exert control over the individual and they are precisely external to ones perception.

On a tangible note, history of gadgets in the field of computer can be traced back to the 17th century. Since last three centuries, the sway of the market has subsumed the individuals to an extent that individuals find themselves almost scrambled when it comes to updating oneself with the new technology which is introduced in the market at the breakneck speed. Just as material social fact affects individual's decision making, similarly The freakish obsession amongst individuals to keep informed about every iota of change in the new gadget is resulting into the formation of new public sphere of dream like fantasies. Such an unprecedented drive and that to an unparalleled scale as compared to history of consumption has supplanted the concepts of subjectivity with reference to time and place. With respect to the intangible orientation of cyber space, like nonmaterial social fact, communications in the digital sphere, along with Digital Games, Digital Music, Digital Research, Digital Security, Digital Videos has turned cyber space into a new paradigm with respect to infotainment as well as models of governance. When it comes to accessing Games, Music and videos on the internet, individuals spend large chunk of their time surfing them. To stretch it further on a quantative note, in 2010 the penetration % of the internet in the Middle East was around 29.8%. The average number of videos watched on You Tube is 2 Billion which has been surpassed by face book with a number of videos of 2.5 billion a day. The number of photos uploaded on flickr in one month is around 130 million. By 2010 the total users on twitter were 175 million. Apart from infotainment, internet is used as one of the potential medium in the areas of research and national security. The two tertiary activities have transformed the cyber space into new forms of governance grid. Apart from developed countries, in most of the developing countries this virtual tool is insistently being used in connecting and finding common platforms between citizens and governments and governments of various countries invest heavily when it comes to national cyber security. Both tangible and intangible domains of internet, like material and nonmaterial social fact impact the decision making, both at individual as well as national level.

As innovation is the key driver behind the inception of cyberspace, in the last few decades it has assumed a cardinal significance when it comes to its application in diverse fields of business expansion and governance operations. These expansions and operations had dissimilar orientations and have ranged to individual profits to societal controls and have finally resulted into possession over information. Both societal forces and technical paradigms, when synergized facilitated in the production of new knowledge. This knowledge intern brought changes in the tangible and intangible structures of the internet. Just as sui generis demeanor of society, internet has also witnessed an organic growth. The point which seeks attention is, it is parallel to the social fact, and it also manifests externality to an individual. Though it has undergone a series of up gradations, it is immune to the existence of a single person's consciousness. To detail it further, an individual as a single unit may or may not use internet but it will still keep on existing. In the same way, the gadgets for accessing virtual space have kept on changing since its inception, due to innovations but it had no impact on its fundamental macrocosm.

Though the social facts are external to an individual but still they obligate certain level of coercion to regulate ones behavior. There is symbiotic relation between a social fact and the coercion posed by it. The level of coercion varies with respect to different point of time and different situation. When social facts thrust various constraints on an individual, they are generically accepted. These constraints endeavor to furnish both societal good and societal duties (Durkheim Emile, 1984: 47). They are intrinsic and prerequisite in conferring morality in society. When it comes to conceiving constraints posed by social facts, we as individuals don't merely imbibe them as they come to us, rather each individual locate his own comfort zone before absorbing them. But still each one mandatorily has to affirm to minimal level of generosity so that the line is not crossed and one may project her or himself as a deviant actor. As in this academic probe cyber world has been juxtaposed to Durkheim's social fact. We can precisely deduce how cyber world also regulate our behavior. As in reality we as individuals while in society confirm to the standards and ethics of our immediate surroundings. Similarly despite the fact that we are using internet, one has to abide by the cyber laws. Using of the virtual space for generating collective conscience can be rated as normal social fact if it asserts the collective willingness of the people for doing something positive. If we access the cyberspace judiciously, such a practice falls under cyber ethics. These ethics entail into moral bindings which refrains an individual from indulging into illicit practices. An individual may surf the internet as per her or his own convenience but should not in any proximity infringe the legal and moral guidelines erected for cyber security. Two of the most serious cyber crimes about which cyber laws restrain and coerce are of piracy and plagiarism. Abstaining from such sort of crimes not only generate cyber trust amongst the virtual communities but also facilitate in ascertaining cyber duties and cyber resilience by taking note of its diverse stakeholders.

As e-world cater to the needs of diverse stakeholders, its penetration in the socioeconomic fabric of the society requires no intellectual comprehension. As in today's world internet is being used at an unprecedented level in each sector of economy in particular and society in general, it has produced a multiplier effect when it comes to growth of economy and creation of jobs as it is a backbone for various manufacturing as well as tertiary sectors of the economy and has several back and forth linkages, employing millions. It would be naive to reduce today's economic interactions to merely a series of tasks with or without external support, rather the contemporary conception of economic interactions along with ICT involve progressive learning and cooperative skills. It is not confined to one single stakeholder but necessitates coordinated efforts of multiple proficient actors who are extremely competent in their own field. Such a scenario fanned by cyber world in the socioeconomic interactions is necessarily replicated by Durkheim's conception of "division of labor". According to this proposal, the phenomenon of division of labor occurs due to social differentiation, which begins with the disintegration of preexisting scenario. It is followed by specialization in occupation which further entails into diversification and multiplication in the industrial activities. To stretch it on a Durkheiman note, the economic actors involved in a legal come economic agreement to increase the output of the collectivity, voluntarily divide the task amongst themselves before the social differentiation (Durkheim Emile, 1984: XIX). As mentioned about the penetration and potential of the internet, with its help, the social, commercial and business spheres have augmented to an unparalleled scale. They are primarily informationally driven. An apt example would be of a BPO industry. as during its inception, this industry primarily dealt in with the information essential for business sustainability but gradually as tertiary sector had been a witness to diversity in their business activities, the entrepreneurs, to manage the business operations in a systematic demeanor has further divided it into KPOs (knowledge processed outsourcing)and MPOs (media processed outsourcing). Such a split is an outcome of social differentiation in the business sector. Few decades back, in India the records in the government offices were kept in black and white, with the advent of ICT, it has become possible for various government agencies to synergize their resources and deliver their services in an effective manner to the citizens. An apt example of that would be the delivering of UID cards and E-payment to government employees.

Durkheim further explains the division of labor, a social phenomenon in the form of three social factors, the volume density, the material density and the moral density of the society. Where volume density stands for size of the population, material density for the number of the individuals on the given ground surface and the moral density locates the intensity of the communication amongst individuals (Durkheim Emile, 1984: 201, 203, 225). All the three social factors in latter and spirit are manifested by social networking sites. The way, in which it has become endemic across various societies, it has ensued into virtual cyber communities. It has entailed into a huge abstract megalopolis without a center. The volume density in the cyber world refers to the total number of internet users across the globe. Material density stands for the total number of users accessing internet or a particular website at a given time and moral density confines to the frequency of internet usage by an individual or a group of individuals. The swiftness and the velocity by which cyber grid transforms and mediates the information and propels the socioeconomic nature of interactions, the cyber fraternity is reliving the age old myth of a flat earth as it has diluted the tangible specifics of time, space and architecture. As people at unanticipated levels become a part of the cyber communities, the concepts like volume density, material density and moral density stand as a paradox when it comes to real and virtual world. On a virtual note, the world is certainly becoming a global village but in reality if not talking about society, even in the families, the personal cohesion amongst members is snapping due to prolonged access to cyber space. So it is not arduous to discern that there is augmentation in the volume density, material density and moral density in a virtual sphere on the other hand there is abatement in material as well as moral density on the real plane.

The ubiquitous demeanor of cyberspace is self-evident. As this e-phenomenon in its sway has reduced the globe into few inches, it has entailed into a double edged sword. On one hand it reciprocates the openness and the flexibility consorted with user empowered ecology while on the other hand it confines to an institutionally-based set of programs that harness the power of technology and control its use and distribution. As deviance had been a innate attribute of human society, given the stature of cyber space, it is not immune to aberrant patterns. We have already discussed that an individual to an extent is free in surfing the internet as per her or his own choice but given the constraints posed by it, one is minimally expected to fall within the broad ethical dimensions. But in many cases users affirm to deviant practices in the lew of short termed gains and end up committing cybercrimes. "Cyber crime is a term used for the attacks done on the cyber security infrastructure of a business organization" (Gottschalk Petter, 2010: 10). Such an malicious attempt can have several goals including cyber espionage which involves illicit access to the target's sensitive information. Ones the access is ensured, the cyber attacker can take full command of the situation and make the organizational information vulnerable as per his own choice. He can extract economic benefits by underselling the stolen information or can delete, modify it causing potential harm to the targeted stakeholders. Apart from this, cyber crime ensuing into cyber anomie involves several other forbidden activities like cyber warfare and cyber terrorism etc (IDSA Task Force Report, 2012: 21). The former is the malignant activities in the cyber space through which the deviant actors attack the information systems of their targets and latter one is an activity in which cyber space is used as a platform to plan terrorist activities. The planning involves recruitment of sympathizers, command and control of propaganda, training of young recruits and raising funding for carrying out terrorists plans. The word "anomie" mentioned in the phenomenon of "cyber anomie" is essentially a Durkheiman conception indicative of normlessness. Such state of affairs is an outcome of abrupt transition in which there is deregulation of normative structure. Such sort of sudden change sabotages the societal structure. When due to failure of instantaneous restoration system, the process of normalization takes time to organize the original structure, during the vacuumed, when the organizational forces attempt to regulate the crisis without any legitimate firm base; it further makes an arrangement fragile (Durkheim Emile, 1951: 15). Similarly when cyber space is attacked by anonymous entities, as mentioned above, the scenario leads to the state of cyber anomie.

As the activities like cyber espionage, cyber warfare and cyber terrorism entails into cyber anomie, sudden upsurge in the revenge porn have preceded into anomalistic dispositions into cyber space. Revenge porn, which was sweeping across the developed world, with the sudden boom with reference to the breakneck access to the E-gadgets amongst the middle class is becoming endemic not even in the metro poles but also in the remote pockets of the globe. It is a new form of crime in which culprits up lode the obscene pictures of their intimate partners on the pornographic sites. Generally culprits in such cases are the jilted and spurned men and the sole objective to up lode the pictures are to beleaguer and tarnish the reputation of their women partners. Since India is one of the developing countries, if we gage the quantum of revenge porn, as per the reports of National Crimes Record Bureau, from 2012 to 2013, there has been 63 % rise in the cyber offences and 104 % jump with reference to obscene content in the electronic form and 1,203 cases registered and 737 people arrested. The acute fix in such versions of cyber anomie is the profile of the culprits. According to the experts, the perpetrators involved in revenge porn hails from posh backgrounds, hence are not similar to other criminals, thus it obscures the nature of crime and it becomes arduous for the legal authorities to take note of them. Even if the authorities attempts to curve the digital prints released in the cyber space, practically it becomes impossible as similar content becomes available on the mirror websites which are approximately 3000 in number. Such an ripping assault on one hand reiterates on the questions of women's vulnerability and on another hand precedes into cyber voyeurism reproducing cyber anomie.

Durkheim, in his book Suicide, a study in sociology in 1897 describes a phenomenon as collective tendencies which engulf susceptible individuals in their sweep. He in a nutshell elaborates over three types of suicides: egoistic suicide, altruistic suicide and anomic suicide. This extreme form of sociological realism, to Durkheim is governed by the forces of integration and regulation in the society. Hence if the suicide in case of is due to high integration of an individual in the society is altruistic in the case of low integration amongst individuals, the suicide is egoistic. In the case of low regulation in the society, the individual is most likely to commit anomic suicide. How cyber world as a social fact integrate and regulate an individual in society needs no introduction. To detail it further, though internet does not plays a direct role when it comes to the relation between individual and his action of committing suicide but recent past there had been several incidents in which internet had been used as a medium in which the act of committing suicide has been shared with a global community. One such prominent incident is of a 19 years old Abraham Biggs who with the help of a life casting website “Justin TV” shared his own live suicide video with approximately 1500 online users and in other case a british man hung himself to death which was also watched live by 100 chat room users. Insanity of such demeanor not only unfurls the spurious connection between a social realism of acute nature but also reiterates the phenomenon of cyber anomie.

On a concluding note, one can deduce that the very essence of the internet is to interrelate the individuals through common bonds. as internet also emboldens the web of relations in the society Durkheim also vents over the notion [society represents itself in a collective form]internet is contributing towards society as well as proving detrimental in one form or the other. So internet is both integrating and disintegrating magnetism for the society. Internet, an opaque phenomenon is a ground reality of the day. So would it be wrong to articulate internet as equal to Durkheim social fact?? Isn’t internet acting as an independent reality? Isn’t it possible to gauge the effects of internet under normal and pathological effects? Isn’t internet external to the individual’s conscience? Isn’t it entails into division of labor. So certainly internet can be admeasure to Durkheim’s social fact. The phenomenon of cyber activism completely aligns with Durkheim’s contemplation of the functioning of the society. Cyber activism has certainly amplified Durkheim’s social thought.

References

Baggio, Bobbe & Beldarrain, Yoany, 2011, *Anonymity and Learning in Digitally Mediated Communications: Authenticity and Trust in Cyber Education*, Hershey, Information Science Reference.

Castells, Manuel, 2010, *The Rise of the Network Society*, London, John Wiley & Sons, Ltd., Publication.

Coiarik, M. Andrew, 2006, *Cyber Terrorism*, Hershey, Idea Group Publishing.

Durkheim, Emile, 1951, *Suicide, a study in sociology*, Glencoe (Illinois), The free press.

Durkheim, Emile, 1982, *Rules Of Sociological Methods*, New York, The free press.

Durkheim, Emile, 1984, *The Division of Labor in Society*, London, The Macmillan Press Ltd.

Ferro, Enrico, 2010, *Handbook of Research on Overcoming Digital Divides: Constructing an Equitable and Competitive Information Society*, Hershey, Information Science Reference (an imprint of IGI Global).

Gelbstein, Eduardo, 2013, Good Digital Hygiene: A guide to staying secure in cyberspace, Webster University, bookbon.com.

Gajaweera, Nalika, 2006, *Bangalore – A Global City? Virtual Realities and Consumer Identities: A research study that unravels the urban experience of Bangalore against the background of its outsourcing boom and corresponding impacts*, Karnataka State Networking Program EQUATIONS.

Gottschalk, Petter, 2010, *Policing Cyber Crime*, Petter Gottschalk & Ventus Publishing.

IDSA Task Force Report, march 2012, *India's Cyber Security, Challenge*, n° 1, Development Enclave, Rao Tula Ram Marg, Delhi Cantt. New Delhi, Institute For Defense Studies & Analyses.

Mitra, Ananda, 1960, *Digital Communications: From E-mail to the Cyber Community*, New York, InfoBase Publishing.

Boyer, Christine, 1996, *Cybercities: Visual Perception in the Age of Electronic communication*, New York, Princeton Architectural Press.

Ritzer, George, 2011, *Sociological theory: eighth edition*, University of Maryland, McGraw-Hill press.

Pastor-Satorras, Romualdo and Vespignani Alessandro, 2004, *Evolution and structure of the internet: A Statistical Physics Approach*, New York, Cambridge University Press.

Butler, Sean, 2000, *The Evolution of Internet interconnections*, Thesis, Rochester Institute of Technology.

Subuddhi, K, 2002 "What Can ICTs do for Rural Poor?", in *Information Technology, Education and Society*, Vol. 3, n° 1/2, James Nicholas Publishers, pp. 47–69.

Thompson, Paul, *Teenager commits suicide live online while 1,500 people watch video stream.* Retrieved from <https://www.google.co.in/search?hl=en-IN&q=Thompson+Paul.+Teenager+commits+suicide+live+online+while+1,500+people+watch+video+stream+%7C+Mail+Online&gbv=2&sa=X&as_q=&nfpr=&spell=1&ei=zYpxU87LFo2iugTYrYDgBg&ved=0CBoQvwU>.

Varuna, Verma, I'm going to ruin you, dear. Retrieved from <http://www.telegraphindia.com/1140803/jsp/7days/18682133.jsp>.

Wilhelm, G. Anthony, 2004, *Digital Nation, Toward an Inclusive Information Society*, London, The MIT Press Cambridge.

Education, Identity and Electronic Communication

ALAIN-PHILIPPE DURAND & ELYSE PETIT

The Teaching of French and Digital Humanities

Upon the publication of *Un Monde techno. Nouveaux espaces électroniques dans le roman français des années 1980 et 1990*, a book that discusses several contemporary French and Belgian novelists, including Jean-Philippe Toussaint, Alain-Philippe Durand began to collaborate with the Belgian writer, photographer, and filmmaker. Between 2000 and 2010 at the University of Rhode Island and since 2010 at the University of Arizona, Toussaint participated in several activities in Durand's seminars on the contemporary novel and film: dialogues with students on the Internet, blogs, on Skype, or face to face during transatlantic visits. Encouraged by these experiences, Toussaint invited Durand to partner with him on what could be a unique project in contemporary literature: to coordinate and maintain in an interactive way, along with his students, the American page of the Jean-Philippe Toussaint web site (jptoussaint.com). Toussaint is one of the rare authors, perhaps the only one, to give an almost unlimited access to his work on the Internet. In this chapter, we give an overview of this rich collaboration and we show how it contributes to the reception, contextualization, and elaboration of Toussaint's work.

The first interaction took place in 2001 when Durand discovered this prophetic Michel Foucault's quotation on the role of space in contemporary society:

> We are in the epoch of simultaneity; we are in the epoch of juxtaposition, the epoch of the near and far, of the side-by-side, of the dispersed. We are at a moment [...] when our experience of the world is less that of a long life developing through time than that of a network that connects points and intersects with its own skein (22).

Durand concluded that the ultimate interaction in his seminars on the contemporary novel would be to go beyond a dialogue that we will describe as horizontal – that is, students analyzing novels among themselves and with the professor; and to add a vertical component – that is, students discussing the assigned novels directly with their authors. Throughout the 2000's, Durand taught seminars in which participated virtually several novelists such as François Bon, Frédéric Beigbeder, Marie Darrieussecq, Benoît Duteurtre, Annie Ernaux, Anne Garréta, François Maspero, and of course, Jean-Philippe Toussaint.[1] These first experiences were successful: American students resumed an interest for the studying of literature and the novelists enjoyed the (rare) opportunity to dialogue in depth with young foreign readers. Consequently, after he found out that Marie Darrieussecq had allowed Durand to create and maintain a web site dedicated to her works (darrieussecq.arizona.edu), Toussaint decided to push the experience even further.

In 2008, in collaboration with computer engineer Patrick Soquet, Toussaint launched a new web site dedicated to his works that would be as exhaustive and interactive as possible (jptoussaint.com). The idea was to propose on the site several geographic zones that represent all the countries and languages where Toussaint's works are translated. For each zone, Toussaint has invited some of his translators or university professors and students researching his works. This is how, along with his students, Alain-Philippe Durand ended up in charge of the Anglo-American zone entitled "Illinois," in homage to the American state where Toussaint's American publisher, Darkey Archive Press, is located. Unlike the vast majority of web sites dedicated to writers, that usually only provide general and practical information (such as forthcoming lectures, bibliographies, media portfolio, biography, etc.), Toussaint was interested in doing something totally different. On his site, Toussaint publishes online, accessible by everyone and anyone, a maximum of original documents and data: outlines, all drafts of his manuscripts, research notes (for instance, in the section on the novel *La Vérité sur Marie*, one can find the entire correspondence that Toussaint exchanged with an Air France pilot). Toussaint also posts audio interviews, personal short films (including some family videos).

In addition, Toussaint gives total freedom to his collaborators to develop the site's geographic zones as they wish. In Durand's case, it was crucial to keep an interactive cooperation that would involve students. Consequently, although the American zone features traditional pages (biography, bibliography, media links, etc.), it also includes a page that features original creations by students that were inspired by their exposition to Toussaint's works. For example, one student preparing a double degree in French and Film Media at the University of Rhode Island, produced and directed a short film on Toussaint's visit to the university in 2010. At the University of Arizona, in Durand's class on Digital Humanities, the syllabus highlights

1 See for example: <http://www.network54.com/Forum/377876/>.

the following tasks that students have to accomplish: "Work along with Professor Durand and Belgian novelist, filmmaker and photographer Jean-Philippe Toussaint. Students will learn about and will conduct research for Toussaint's official web site. In addition to applying research techniques to updating and publishing on the sites, students will create original works (creative writing, films, articles, interviews, photos, paintings or any other work of art) based on their reading of Toussaint's works." Nevertheless, in spite of these activities' success and the high interest that students showed for Toussaint's original documents posted on the web site, this is only the beginning of what is known today as "digital humanities."

Before we proceed further, it is important to present a brief inventory of digital humanities. Whoever is interested in learning more about digital humanities should consult the American journal *Digital Humanities Quarterly*, published by the Alliance of Digital Humanities Organizations since 2007. We especially recommend Patrick Svensson's seminal article "The Landscape of Digital Humanities." Reading it, one understands how difficult it is to define what digital humanities are exactly. In his article, Svensson provides some of the main proposed definitions of digital humanities. All of them underline the still emerging status of this discipline. According to the journal *DHQ*, digital humanities is "a diverse and still emerging field that encompasses the practice of humanities research in and through information technology, and the exploration of how the humanities may evolve through their engagement with technology, media, and computational methods." For the Digital Humanities Manifesto 2.0, digital humanities "is not a unified field but *an array of convergent practices* that explore a universe in which: a) print is no longer the exclusive or the normative medium in which knowledge is produced and/or disseminated; instead, print finds itself absorbed into new, multimedia configurations; and b) digital tools, techniques, and media have altered the production and dissemination of knowledge in the arts, human and social sciences." Finally, according to N. Katherine Hayles, "it would seem tenable to state that minimally, digital humanities is manifested by a single scholar, teacher, artist, programmer, engineer or student doing some kind of work – thinking, reflecting, writing, creating – at the intersection of the humanities and information technology – or by "products" resulting from such activities."

At the University of Arizona, several of us are conducting interdisciplinary research projects in digital humanities that involve three colleges: the College of Humanities, the College of Science, and the College of

Social and Behavioral Sciences. For example, Jiang Wu, an Associate Professor of Chinese, works on the spatial analysis of religions: Wu builds databases on Buddhist manuscripts and monasteries in south-east Asia (Buddhist Geographic Information System – BGIS). We are facing the same situation as Jiang Wu: we are collecting, organizing, and archiving immense databases that are now simultaneously available and accessible worldwide in just one click. But the following question remains: what to do with this wealth of material? How to maximize its huge potential? How to go beyond archiving and informing and establishing new practices, new creations? These are the questions Durand decided to ask Elyse Petit, one of his doctoral students in Second Language Acquisition and Teaching. Petit is conducting her doctoral research on the use of new technologies in the teaching of French and francophone literature and film.

In 2012, when Durand asked Petit to work on the web site of the two francophone writers J.P Toussaint and Marie Darieussecq, she saw a great opportunity to combine her research interests of French literature, technology and pedagogy. Her first objective was to redesign and update both sites. While she was exploring and rearranging the Darrieussecq's site, she realized the richness of the multimodal texts provided. As explained previously, these are not just general and practical informative and promotional sites about francophone authors, but rather they provide various and unique multimodal texts such as manuscripts, exclusive interviews, short movies, audio and video clips, images, and so on. Petit proposed to Durand to go further and to develop lesson plans which could be implemented in a foreign language classroom, and eventually could be posted on both sites in order to offer pedagogical activities as well as a way to share such activities and ideas with other foreign language instructors around the world.

Petit created two different lesson plans that are used as examples in this chapter: one based on the Darrieussecq's site and the other based on J.P Toussaint's. First, it seems important to explain the historical aspect and the pedagogical framework on which the development of these lesson plans are based, that is "multiliteracies pedagogy." Traditionally, the concept of literacy is understood as the ability of a person to read and write texts. Thus, anyone who had access to formal education is considered literate. However, globalization and the evolution of technology and multimedia have recently shift the way literacy is defined. Decoding and encoding texts are no longer considered sufficient. Rather, it is fundamental to develop the skills of reading and writing in order to interact with others, and to become socially, economically and culturally integrated into the world.

In the 1970's and 1980's the sociocultural view proposed to redefine literacy as a social practice embedded in societies. Indeed, the evolution of the world provided a multiplicity of media and contexts, impacting the notion of literacy which evolved into plural practices. Therefore, it also became necessary to gain the ability of encoding and decoding semiotic resources such as advertisements, road signs, logos, smileys, text messages, Facebook and Twitter posts, etc. The concept of literacies was extended in many academic fields. In the early 1990's, various studies and researches focused on concepts such as "media-literacy education," which examines the learning of reading semiotics images, or "e-literacy," which implies the use of technology and digital texts. In 1996, the term of "Multiliteracies Pedagogy" referred to the shift made to the notion of literacy in the domain of education. Although elements of multiliteracies-based approaches to foreign language teaching can be found as early as 1991 in Swaffar, Arens & Byrnes, the roots of multiliteracies can be traced to the 1996 New London Group's (NLG) article, "A Pedagogy of Multilteracies: Designing Social Futures." In the early 1990's a group of ten scholars decided to meet in order to discuss the shift that occurred in literacy pedagogy due to changes in communication media and to cultural differences. In 1996, they wrote this long article-manifesto which "is a theoretical overview of the current social context of learning and the consequences of social changes for the content (the 'what') and the form ('the How') of literacy pedagogy" (NLG 1996: 4).

In May 2007, a report by the Modern Language Association (MLA) opened new perspectives in researching the teaching of foreign languages. Language is no longer conceived as a communicative competence, but its use is also recognized as a social act (Gee 1996). Language indicates the way people perceive themselves, others and the world, and the way they interact with others: "Language is a complex multifunctional phenomenon that links an individual to communities, and to national cultures." (MLA Ad Hoc Committee on Foreign Languages 2007). Consequently, education scholars (Kramsch, Swaffar, Knobel & Lankshear, 2006) analyzed the methods that were used in foreign language acquisition and teaching. They question oral proficiency which was predominant in language teaching, in favor of the reading and writing skills which they started raising as central in order to improve learners' communicative competence and discourse ability. The multiple modes of communication and the cultural and linguistic diversity included in texts, require researchers to propose the development of a pedagogy that takes into account the plural literacy in language acquisition. A literacy that would include linguistic, cognitive and sociocultural dimensions according to Kern (2000).

The pedagogy of multiliteracies presented by the New London Group develops a different kind of pedagogy, one in which language and other modes of meanings are dynamic representational resources constantly being remade by their users as they work to achieve various cultural purposes. As Kramsch advocates "today, it is no longer enough for learners to be able to accurately construct and understand sentences to convey or receive information; they have to develop the ability to understand, interpret, and produce a variety of multimodal texts in the target language, and understand how language and other symbolic systems are used in order to design meaning" (2006: 251). The "design of meaning" (the "what") is a key element of literacy-based instruction and is represented by three principles: available designs, designing and the redesigned.

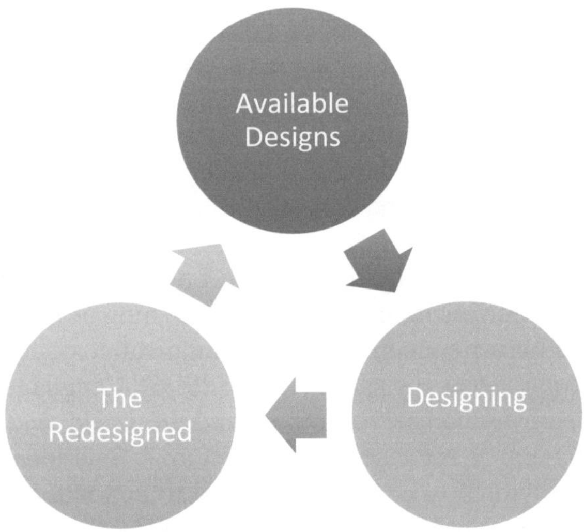

Available Designs are linguistic, social and cultural resources that learners use to interpret or create texts.

Designing is an active, dynamic process that involves the creation of form-meaning connections through interpretation or creation of texts.

The result of Designing is a transformed representation of ***Available Designs*** called the ***Redesigned*** (which become new available designs).

In order to carry out a pedagogy of "design" and "multimodality" and facilitate learner engagement in acts of meaning design, NLG proposed the paradigm of the Four Curricular Components as a way to frame activities in and out of class. These components – (called situated practice, overt instruction, critical framing, and transformed practice) – involve interpretive and productive engagement with texts, and metacognitive reflections on how meaning is made as well activities which draw attention to how form (linguistic, spatial layout, color, etc.) creates certain meanings and not others.

Situated practice has been referred to as experiencing, or "immersion in the text." It must consider the affective and sociocultural need and identities of all students and must constitute an arena in which learners are secure in taking risks and trusting the guidance of others (peers and teachers). Some of the things learners might do in this phase are describe, explore, observe, and connect with a text.

Overt instruction is really about drawing attention to how form (linguistic, spatial layout, color, etc.) creates certain meanings and not others. So this might involve learners reflecting on why an author, for instance, might have chosen to use the present tense to narrate a story that took place in the past, rather than the past tense. The goal of Overt Instruction is to gain a consciousness and to take over what is being learned.

Critical framing involves metacognitive reflections on how meaning is made. Analyses of author purpose, of ideologies behind texts. Learners can achieve the necessary personal and theoretical distance from what they have learned. Activities in this phase might involve learners in comparing, synthesizing, critiquing, and understanding.

Transformed practice consists of activities which allow learners to apply what they have learned. In this phase, learners might compose, construct, design, and demonstrate. They try to re-create a discourse by engaging in it their own real purposes.

Taken together, the four curricular components form a paradigm which can be helpful for the structuring of a daily lesson plan. There is no pre-given sequence that must be followed, rather these can be interwoven in a non-linear, recursive sequence, and many activities might simultaneously involve more than one of these curricular components. The lesson plan that Petit has developed for J.P. Toussaint's website is based on the framework of multiliteracies, and presented under a personal format of lesson plan created for her own courses. Another framework used in order to develop the project zooms out from the daily activities and offers a way to structure the content of an entire course, or even an entire curriculum, such as the Georgetown University's department of German Studies. The genre-based instruction proposed in this framework involves the notions of genre not only as structural forms, but socially constructed, purposeful texts. The lesson plan presented here is not definitive and can be reorganized or modified in order to aim students' level and teacher's objectives. It implies that students are able to read the novel (which can be provided in English). The lesson is not made to ask students to read an entire novel in the target language which can be too difficult depending on the level, but to develop learners' awareness of genres and foster a first interest in francophone literature and contemporary writers. The objective of this lesson offers an overview of a pedagogical approach that could be used in any proficiency level in order to develop students' meaning-making and critical-thinking. First, it demonstrates that literature can be implemented in foreign language classroom as soon as the first level of learning with scaffolding and appropriate activities. Then, it presents some activities that can be designed from an authentic and digital text in order to develop students' strategies for reading the new and unfamiliar.

As indicated earlier, the four curricular components are interchangeable and are not made to be performed in any specific order. They can be repeated several times in a single lesson, it depends on the purpose of the lesson and the goals and objectives. In the lesson plan that Petit proposes, the four components are exposed fairly orderly but only to signify the types of activities available for each component. By offering different activities the instructor can then choose to develop and construct her/his own lesson plan. The development of a lesson plan is to arrange various activities that allow students to achieve pre-determined goals to improve their language skills and their cultural awareness of the target language. The activities presented are only samples of what can be drawn from the novelist's web site. They could be implemented within a broader package about francophone literature, or a work of creative writing, or on the notion of genre.

Jean-Philippe Toussaint, *Faire l'amour* (Minuit, 2002)
Lesson Plan: Part 1

Course Topic
Genre-based instruction

Resource
Web site: Jean Philippe Toussaint <http://jptoussaint.com/faire-l-amour.html>

Language proficiency level
Fourth or fifth semester of French

Lesson
This lesson could be divided in two parts (from 1:30 to 2:00 hours each)

Homework to prepare the lesson
➢ Blog :
– Read J.P. Toussaint's novel *Faire l'Amour*, give a summary and write a personal reflection.
– Comment peers' blogs

Lesson objective
Develop students' awareness on different genres of texts.

Goals
From the reading of the novel, students will be able to interpret an excerpt of the novel presented from different multimodal texts (audio, video, and printed texts).
Students will be able to identify and recognize the different genres and their structures.

Objectives: Students will be able to:
- Analyze different structures and conventions of an authentic text.
- Identify and recognize different lexico-grammatical parts of sentences or paragraphs.
- Think critically about the cultural differences presented in the diverse texts.
- Discuss and argument in the target language.
- Identify stylistic choices made linguistically in the different texts.

Activities
➢ Situated practice:
– Ask students to explore the different covers which represent the novel. What do they see? Which ones are the most representative of the plot? Explain?
– Ask students to base their reflection on the concepts of cultural differences, audience, and intention of the author.
➢ Critical framing:
– In groups, students will imagine they are the author of the novel and decide on an illustration they would choose for the cover.
– Write a short paragraph to argument and report to the class orally.

➢ Situated practice: – From the web site, project "la première ébauche" (2000) of the novel, and ask students to read the first paragraph. Then ask students to discuss in pair the following questions: Who? When? Where? What? Why? – What does this excerpt tell about the plot? The characters? And the action? ➢ Overt instruction: – Based on the out-of-class reading, from which moment in the story does this excerpt come from? Ask students to collect vocabulary/expressions that demonstrate their thought. – Students listen to the same excerpt read by the author (text audio provided on the site) and see if there are differences between both texts. – From both texts given to students, underline the differences between the "Ebauche" and the final version of the text. ➢ Critical framing: – Discuss in groups the differences and why does the author make these stylistic choices? What do the transformations create to the text? The audience? – Report to the class. ➢ Transformed Practice: – Based on the excerpt, students imagine that they are in possession of the flask. Using the style, the language register, the intonation, the tense of verbs and the syntax chosen by the author, they should write a paragraph of what they could do with it.

Part 2

Homework (previous day) ➢ Blog : – From the link "Films: Faire l'amour, une lecture japonaise," students will read and explore the presentation and the "journal du Tournage." Then they should write a personal reflection. – Comment peers' blogs.
Course Topic The notion of genre
Goals From the reading of the novel, students will be able to interpret an excerpt of the novel presented from different multimodal texts (audio, video and printed texts). Students will be able to identify and recognize the different genres and their structures.
Objectives: Students will be able to: • Collect features of different genres presented. • Compare and contrast the different multimodal supports of an authentic text. • Analyze and identify the relationship between text and images. • Create a technical template from a model.

Activities

- ➢ Situated practice:
  - To create schemata activation, ask students to summarize orally what they have learned from the previous class: what is a *genre*? Give different genre with which you are familiarized in your daily life.
- ➢ Overt instruction:
  - From the previous class, as well as the readings that students have done for today's homework, in groups, students should create a document that describes the different features of each genre. Teacher could create a Google drive document for each group that represents each genre. At the end of the class, each group could share their own document with the rest of the class.
- ➢ Situated practice:
  - Using quotes from students' homework blog, teacher ask student to reflect orally, argument and justify their thoughts.
- ➢ Overt instruction:
  - From the features of the different genres identified by students, teacher asks students to predict how Toussaint would film the text. They should explain using the text itself.
- ➢ Critical framing:
  - In small groups, students fill a technical form in which they will provide details on their vision of the film they want to direct and record based on the novel: actors' choice, place, time, colors, camera angles, language register, and so on. Instructor can provide "cinema vocabulary" the students could use in order to fill the form.
- ➢ Situated practice:
  - With the whole class, the instructor presents the first excerpt of the film available on the site, and asks students to react: are you surprised? Do you like it? Do you situate it from the novel? Which genre is it?
- ➢ Overt instruction:
  - In small groups, students compare and contrast the excerpt from a genre perspective: What does the written text bring compared to the audio text and to the video? Explain with examples from the texts.
- ➢ Transformed Practice:

<u>In class</u>:
From the same excerpt, write it as a dialogue: students can choose the style and the register, but they should keep the same plot.

<u>Outside class (as a final group project)</u>:
- Students present a technical form of the sequence (place, actors, costumes).
- Students must create their own video and post it on the blog.
- Students must write a "journal de tournage" using Toussaint's model.

In August 2014, Toussaint invited Petit to travel to Brussels for a workshop with him and his collaborators to develop an interactive writing project, "the Borges Project," for his web site:

In *La Vérité sur Marie*, Jean-Philippe Toussaint evokes a short story by Jorge Luis Borges, "The Island of Anamorphoses." He writes: 'The Island of Anamorphoses, Borge's apocryphal short story in which, depressed and beaten at the end of a long process of solipsist decline, the writer who invented the third person in literature ends up abandoning his invention and goes back to writing in the first person.'

All traces of this captivating short story seems to have disappeared. You're a writer, professor, student or book lover? Why not rewrite it? Or imagine what became of it? We'd like you to send us your own version of this missing story.' (Retrieved from <http://www.jptoussaint.com/borges-projet.html>)

The workshop lasted three days during which Toussaint first asked Petit and his collaborators to read and edit selected texts submitted to the "Borges Project" from all over the world. Selected texts were then published onto Toussaint's website. Also, the workshop consisted to standardize the pages among the different languages provided on the web site. Finally, brainstorm reflections and discussions brought new ideas to implement on the site.

Working in collaboration with Toussaint and his team on these various interactive projects that involve a writer and his readers demonstrates that new perspectives and approaches are appearing in the writing, reading, and teaching of contemporary literature, as well as in the teaching of French or other second languages. The developing field of digital humanities offers endless possibilities. Consequently, Durand and Petit envision future research projects that would look at the way digital humanities directly affect and change the ways novelists elaborate their works and how they interact with their readers/viewers.

References

Allen, Heather Willis, & Paesani, Kate, 2010, "Exploring the Feasibility of a Pedagogy of Multiliteracies in Introductory Foreign Language Courses," in *L2 Journal*, n° 2, pp. 119–142.

Cope, Bill, & Kalantzis, Mary, 2009, "Multiliteracies", in *New Literacies, New Learning. Pedagogies: An International Journal*, n° 4, pp. 164–195.

Durand, Alain-Philippe, 2004, *Un Monde Techno. Nouveaux espaces électroniques dans le roman français des années 1980 et 1990*, Préface de Marc Augé, Berlin, Weidler.

Foucault, Michel, 1986, "Of Other Spaces", in *Diacritics,* 16.1, pp. 22–27.

Gee, James, 2011, *Social Linguistics and Literacies: Ideology in Discourses*, 4th edition, London, Routledge.

Hall, Joan Kelly, 2002, *Methods for Teaching Foreign Languages: Creating a Community of Learners in the Classroom*, Colombus, OH, Merrill Prentice Hall.

Kern, Richard, 2000, *Literacy and Language Teaching*, Oxford, Oxford University Press.

Kramsch, Claire, 2006, "From Communicative Competence to Symbolic Competence," in *Modern Language Journal*, n° 90, pp. 249–252.

MLA Ad Hoc Committee on Foreign Languages, 2007, "Foreign Languages and Higher Education: New Structures for a Changed World," in *The Modern language Association*.

New London Group, 1996, "A Pedagogy of Multiliteracies: Designing Social Future's," in *Harvard Educational Review*, 66.1, pp. 60–92.

Svensson, Patrik, 2010, "The Landscape of Digital Humanities", in *Digital Humanities Quarterly,* 4.1. Retrieved from <http://digitalhumanities.org/dhq/vol/4/1/000080/000080.html>.

Swaffar, Janet, & Arens, Karen, 2005, "Remapping the Foreign Language Curriculum: An Approach Through Multiple Literacies," New York, Modern Language Association.

Toussaint, Jean-Philippe, 2002, *Faire l'amour*, Paris, Minuit.

Toussaint, Jean-Philippe, 2009, *La Vérité sur Marie*, Paris, Minuit.

Marianna Vivitsou & Kirsi Viitanen

The pedagogies of the future: through young people's eyes in storytelling experiences with the digital in Finland and Greece

In search of digital pedagogies

Social media create spaces where young people produce content, publish image-based artifacts, and participate in text-based discussions. The increasing appearances of video-sharing sites (e.g., You Tube) and photo-sharing sites on the Web and the fact that social networks (e.g., Facebook and Twitter) nowadays support and encourage moving and still image broadcasting and sharing indicates a fundamental shift for human social communication. The stronger focus on visuals gives way to communication artifacts that bring together semiotic systems and codes of movement and change that are not occurring on the basis of a linguistic system, exclusively. In this way, by allowing a different 'language' to come up through image-based practices for movement and change, social media seem to create another space. When social media are used for pedagogy and learning with image-based practices, it is in that new space where pedagogy can transform. The primary question now seems to be whether this kind of transformation is feasible and, if it is, what it entails.

One way to look at pedagogy is to think about its structural components. What variables do we need to consider before we are able to delineate a pedagogical space, an area where individuals gather in order to learn, develop identities and grow? As Simon (1988) argues, a comprehensive view of pedagogy would include the integration of particular curriculum content and design, classroom practices and techniques, time and space for the practice of strategies and techniques, and evaluation purposes and methods. Would that mean that if we manipulate one or many of these variables the result would be a transformed pedagogy? Can, therefore, social media as space where collaboration, communication and learning can occur transform pedagogy? One thing that social media, being itself digital technology, can certainly do is add a new denotation to pedagogy.

Digital pedagogy is a term that we can trace as an attribute of Digital Humanities or a brief for Digital Humanities Pedagogy (e.g., as found in Hirsch 2012). Scholars in Digital Humanities (e.g., Hirsch 2012) argue that pedagogy is neglected in the process of humanizing the integration of the digital in the pedagogical practice and suggest that a lot more than the collaboration in networks of people that live a constantly visible life online (Kirschenbaum, 2010: 55, 60) should be included in the discussion of what constitutes the identity of Digital Humanities (Svensson 2010). The considerations that link to this identity vary from views of information technology as exploration and epistemological critique (Balsamo 2000) to traditional views that approach technology in a more formal and instrumental way (Kirschenbaum 2004). The question, however, of what changes when pedagogies blend with the digital remains. Hirsch (2012) urges us to reflect critically about pedagogy by asking questions about 'what it is that we *do'* (author's emphasis) as digital humanist pedagogues. Drawing from Irvine (2006), Hirsch (2012: 16–7) poses questions about whether we simply teach with the digital or whether teaching with it legitimizes us to declare our scholarly understanding of pedagogy with the digital as being its intersection with computing. Either way, Hirsch seems prone to align with Irvine (2006: 153) in that teaching is one aspect of the process of learning to enact as pedagogue, whether professor or teacher.

The integration, however, of digital technologies in pedagogy is an ideological and political decision. Critical theorist Henry Giroux (1994) when discussing what constitutes essential questions to ask about pedagogy draws the distinction between questions about how students learn and about how educators construct the ideological and political positions from which they speak. Simon (1988) claims that "talk about pedagogy is simultaneously talk about the details of what students and others might talk together and the cultural politics such practices support". To advocate then for a pedagogy with the digital is to advocate for a political vision. "In this perspective", Simon (1988) concludes "we cannot talk about teaching practice without talking about politics".

What, then, can this political vision entail and whose political vision can it be? Drawing again from Digital Humanities literature we align with de Smedt (2002) in the view that "More important than the use of machines" are the "new ways of thinking". For digital humanists these, among others, include thinking with and against the machine in, for instance, teaching research methodologies and introducing programming in the pedagogical practice. The vision, therefore, of pedagogy with focus on technological advanced is a vision incomplete. More aspects are needed for the 'new thinking'. These should entail a culture of collaboration among those involved in digital pedagogies: pedagogues, scholars, dedicated expert programmers and software developers, in addition to other academic and technical specialists (Short 2010). "Specialization and expertise in one discipline comes at the opportunity cost of another", Short (2010) argues and goes on to add that what distinguishes digital humanists from traditional humanists, perhaps, lies in a willingness to embrace collaboration as mode of research. But to what degree are these 'new ways' actually new? Is it indeed at collaboration where we should draw the line between traditional and digital pedagogies? Is the digital where we should draw the line between traditionalists and non-traditionalists? Does developing digital skills make the difference? And what is there beyond the development of the ability to participate in the design and creation of new media (Ramsay 2012)? What role do artifacts of communication play?

Learning how to code indeed takes the development of a skill but the code itself is 'language' that the programmers use to eventually take a position, express a feeling, make a promise, and, therefore, communicate a message ('speech' act, sign). In this sense, the code is an artifact of communication as moving and still image-based digital creations are. As these present the coder's, or creator's worldview, they are also cultural objects amenable to express critique and raise questions, just as political visions do. The 'language' of the code, therefore, and the 'language' of the en-coded artifact can become the vehicles for the creator to express a political vision. By using Wikipedia as an example platform, Kill (2012) claims that such artifact of communication creates the space for students to become active participants in their own digital 'social futures', by integrating an understanding of writing for individual inspiration with insights into writing as an arena of social action. Although Kill's article shifts the focus to a digital pedagogy that aims to shape students into informed, civic-minded, digital participants, the young people seem to have little, if at all, a say in the process of shaping the collaboration, of constructing the artifact or, even, of deciding what they themselves want to learn.

The question therefore of whether, to what degree and how social media and digital technologies transform pedagogy does not seem to suffice;

the question about the extent pedagogy itself becomes transformative through the use of social media should be asked instead. We believe that one way to make this happen is by allowing a culture of inclusiveness to emerge.

It is our position, therefore, that pedagogies with the digital, or without it, cannot be transformative unless the focus shifts so that, in addition to the (digital) pedagogue, it includes the youth as decision makers and as curriculum planners. In other words, in the pedagogies of the future the young people are leaders, partners, and architects of their own learning. In this paper we argue, by discussing storytelling experiences with the digital of young adolescents from Finland and Greece, that such shift is feasible. To support our case, we examine what initiative and action these young people develop as digital storytellers and short filmmakers in collaborative physical and virtual spaces.

The pedagogies of the future

To respond to the question asking what makes the pedagogies of the future transformative we need to reflect critically on how the youth will be enabled to transform their lived experiences into knowledge; and how they can use the already acquired knowledge as a process to create new knowledge, and, in this way, to participate in the dialogical process of learning and knowing (Freire 2005). It seems that, even in the current era where we seek to define pedagogies in terms of the digital, questions about how being and knowing are related are central issues for consideration. Clearly, however, there should be a transposition from pedagogy as teacher-led to one that is youth-driven. Following Freire's idea as discussed in the Pedagogy of the Oppressed (2005: 19), we believe that apprenticeship with the new body of knowledge is one way to keep the dialogue going. How can pedagogies become digital if young people do not experience learning with connective platforms, in social networks, with mobile devices, within institutional premises or other urban spaces? How is the pedagogical dialogue of learning and knowing to keep going without the epistemological curiosity that the experience of connectivity generates? Certainly, pedagogy with social media and networks, as it occurs in multiple connected locations, it also involves multiple participants, such as teachers, experts and colleagues, young people and peers. This new pedagogical setting disrupts the traditional teacher-student situation by bringing in a variety of perceptions of reality. It is a different thing to have to manage solidly structured pedagogical content from the one that is constantly re-shaped, like a story incomplete.

We believe, therefore, that digital pedagogies of the future should create spaces of unpredictability. Rather than perpetuating notions that present knowledge as stable entity, therefore, we should be ready to suspend certainty and re-negotiate with students (Matusov 2009: 5) what, how, where and with whom they want to learn. Suspending the certainty of knowing what is known or what needs to be known and keeping the pedagogical dialogue alive as story incomplete are essential elements of relational pedagogy (Bingham and Sidorkin 2001, Matusov 2009). The pedagogies of the future cannot but open up relational spaces. In such spaces young people have the opportunity to imagine alternative positions of themselves as subjects in the world and figure out what divergent social designs could look like (Freire 2005: 24) on the one hand.

On the other, in such spaces young people should have the opportunity to develop literacies in multiple semiotic domains. The need for understanding visual and multimodal literacies is discussed in Gee (2004). Gee explains that nowadays people need to be literate in many different semiotic domains, and be able to become literate in other new semiotic domains throughout their lives. Jenkins et al (2009) shift the focus of individual literacy development underlying Gee's ideas to the community, where, as they argue (Jenkins et al 2009: xiii), lie the strong drivers for creative expression and active participation.

Digital storytelling is one pedagogical method that could allow for relational spaces where, through connective technologies, young people grow and learn, by, among others, developing multimodal literacies. We call the connective spaces that can foster relational pedagogies social media for pedagogy, or pedagogical social media.

Digital storytelling as pedagogical method

The content of learning in pedagogical social media can be made visible in the form of stories. Stories (Herzog 2001: 185) result from actions and words that become the content of stories. When young people share and tell digital stories in pedagogical social media they enter the process of recounting present conditions in the light of past memories and future anticipations (Arendt 1968, Kearney 2002, Ricoeur 1992). They do so by drawing from lived experience that they need to transform into a meaningful whole in the form of short videos. As young people try to connect the parts together, whether to recount the story of daily routine, a historical event or a lab experiment they need to interpret the event in the present time in the light of what preceded in the past and what will follow in the future. In this way, the teller develops a sense of the self in relation to the story and by answering questions, such as "why and how did the event happen?" and "what will follow?" As the events of the story take their place in the time continuum the self develops a relation with how the plot evolves and, as a result, a coherent narrative identity. The process of creating the story does not change the perception of the self only. Digital storytelling, in addition, changes the way we perceive pedagogical "*time by transforming it from an impersonal passing of fragmented moments into a pattern*" (Kearney 2002: 4) where learning becomes part of the story plot and has present, past and future.

Studies with digital storytelling as background pedagogical method show that it can cater for the representation of the agentive self (Hull and Katz, 2006); the construction of narratives of self, family and community (Hull, Kenney, Marple, & Forsman-Schneider 2006); and for self-presentation (Nelson, Hull & Roche-Smith 2008) and making meaning through symbolic means and subject matters. Other studies discuss storytelling as a narrative literacy that can convey the emotional states of characters and authors by shifting modes of expression (Yang 2012), and as means for marginalized teens to manage the fear of difference in online discussions about the self and others (Snell 2012).

Digital storytelling enhances building knowledge upon the young people's capabilities so they can meet the requirements that the digital era generates in a humanized way. However, it does so in an unpredictable way, as the constantly emergent narratives cannot have a fixed beginning, middle or end. It all depends on the perspective of the teller. When the youth decide to tell a story they also need to take a position. Through the story, therefore, the vision of the teller becomes visible whether this is about citizenship, human relations or scientific achievements. In this sense storytelling is a type of political act that, although it lacks a uniform fashion of expression, does have intentionality. One such intentionality derives from the desire of the teller to share and tell stories online. To get there, the youth need to develop different types of initiative in terms of how they act with peers and the medium of communication (be it language or other system) within the learning space (be it cultural or natural landscape or a connective environment). So much more for unpredictability, therefore, when the story making and telling takes place in a collaborative mode.

In this study, by 'initiative' we mean the "intervention of the agent of action into the course of the world, an intervention that effectively causes changes in the world" (Ricoeur 1992: 109). When the individual's ability to initiate a new story meets with the peers' in a group or a small team there is need for students to re-negotiate what they know already in order to establish relations and work, create and learn together in the physical classroom and the virtual pedagogical space. To get there they also need to change 'who' they are in the present time. Social media and image-sharing sites make this need for change even more pressing nowadays.

As the young people plan clips for the digital story they need to activate filmic practices and shoot, for example, from a particular angle to convey a specific message. Then they need to think of the story as sign or, in other words, they need to think in terms of a system other than language. And this is where the youth develop multimodal literacies. Later, when they come to negotiate with peers what, for instance, the best shots are they need to come up with convincing arguments in order to prove their case. Therefore, as Bingham and Sidorkin (2001) would put it, there are quite a few knowledge 'gaps' for young people to cover in the process of making and telling a meaningful story. These range from gaps at the individual to the collective level. Whatsoever, these gaps are about changes of knowing and being not only for oneself but also with others. Smyth et al. (2013) would possibly argue that digital storytelling fulfills the need to reconsider fixed notions of learning by offering an alternative notion. We believe that growing and learning with peers and building knowledge and relations in communities is by no means fixed notion for pedagogy; nor is for digital pedagogies. Nowadays, however, this can happen in social media spaces.

Social media spaces for pedagogy

So far we have discussed the initiative and action the youth take in the process of making and telling digital stories in order to delineate what we mean by digital relational pedagogies. The question, however, that comes up next is whether there is room for interventions and change in social media and connective technologies and, if there is, what it looks like.

In their study of networked publics, boyd and Marwick (2011) argue that what teens find in popular networking sites is the space for identity work and social conceptualization. These are elements that young people seek in Facebook, Twitter and other spaces, being ostracized from publics (e.g., clubs) that they, nevertheless, wish to enter. When networked environments host stories that draw from pedagogical subject matters, they open up the space for learning in 'public' (boyd & Marwick, 2011). One such social conceptualization is that the youth view themselves as part of the 'public', more like 'citizens' as actively involved individuals rather than students. They do so by acting to make and publish the story on the one hand. On the other a shift in perspective also happens when watching the connected peers' digital stories where they discover alternative patterns of life and society. These they extrapolate later to their own narratives by, for example, transferring the setting of the story from school premises to alternative, urban spaces. In this way, the young filmmakers create their own blueprint of how to make and tell the story. In a sense, they themselves become digital storytelling curriculum planners.

Therefore, the network does not only become a collaborator in the identity formation process as Marwick and boyd argue (2011); nor do social media are only spaces that enable narratives of the agentive self and the community to come up, as we discussed above. In addition, the network takes on the role of mediator (Verbeek 2006) in terms of the ways the young discover the world as cultural landscape where they relate with humans, other species and the environment. We believe that this experience of being in the network as storyteller and filmmaker can create opportunities for initiative. Digital stories are, as Sartre (1967) would put it, 'directed creations', in the sense that, being artwork, they mobilize the power of expression and imagination. By making, telling, publishing and sharing the stories on social networking sites, the young people can develop not only an appreciation of the world as cultural landscape but also an appreciation of the digital story as artifact. This can bring forward new, experimental modes of expression. Either way, by making stories that advocate, for instance, environmental sustainability or promoting a new artistic expression, there is probability that this experience shapes caring individuals who might one day choose to engage "in the communal project of better world making" (Saito 2010). The networking experience, therefore, when based on relational pedagogies can create the space for young people to develop both as media creators and as thinkers. Glimpses of relational pedagogy can be traced in, for example, the notion of participatory culture as used by Jenkins (2012: 7) to describe the situation where there are "relatively low barriers to artistic expression and civic engagement". Jenkins et al. (2009) argue that the value of, for instance, creative writing or art classes does not only lie in the possibility to identify and train future writers and artists, but also in the opportunity that opens up for every young person to express through words, sounds, and images, even if most will never do so professionally. Having these experiences changes the way the youth think about themselves and work created by others (Jenkins et al. 2009).

By pedagogical social media, however, we do not only mean the connective environments where the young people upload and share their digital stories. We also include the technologies that need to act in concert to make the storytelling happen. Such technologies vary from the digital devices that the tellers use to make the story (e.g., digital cameras, tablets, smartphones etc.) to software for editing and finalizing the story (e.g., by adding annotations or subtitles). Given the complexity of the situation, we view pedagogical social media as spaces where different sorts of unpredictability can rise up. Incompatible or outdated software, unstable connectivity and low bandwidth are some examples of disruptive mechanisms. Technologies, therefore, like the Pandora's box of the digital era, package both hopes and fears. One such hope is that they create the space for young people to become independent from the ability of the adults by managing difficulties that the adults are not prepared or trained to. Pedagogical social media therefore are spaces where the young are also the authorities, take decisions and solve problems. These can become spaces of initiative where the young people take action and, by articulating own voices and suspending belief of what is thought to be known, make changes to their world.

To understand how students think when learning with technologies, in this study we examine the types of intervention and the subsequent changes that take place in storytelling pedagogical social media experiences. To do so, we will discuss and analyze Finnish and Greek young adolescent interviews.

The study, the methods and the participants

In this qualitative study we draw from interviews of focus groups of Finnish and Greek students in order to seek answers to our main research question, "What initiative do young people come up with when telling digital stories in pedagogical social media?"

During interviews the students (2 pairs and 3 groups in the Finnish case, N= 14, and six groups of four in the Greek case, N= 24) responded to questions concerning the process of making stories, what problems they met and what actions they took in order to resolve these. Also questions about the key moments of the experience were addressed. Storytelling took place as part of the 'Boundless Classroom' project that connected students and teachers from California, Finland and Greece (October-December 2012) with digital technology-enhanced learning activities. In this study we will discuss the experiences of students from 3 schools in Finland (School 2, 4 and 10) and 3 schools in Greece (School 1, 2 and 4) with a total of 120 participating students (60 in each country).

Considering that the project spans across a variety of pedagogical contexts, participants overall make decisions according to own needs. Therefore, technologies are integrated organically in the school everydayness with the use of digital and mobile technologies for storytelling as common background and the Mobile Video Experience (MoViE) <http://cicero-movie.edu.helsinki.fi> as the web-based platform that connects teachers and students. The teaching plan addresses each and every classroom needs while the research team provides mainly technical support during the implementation of the storytelling activities.

We collected the data while the activities were still running by observing students shooting the scenes, working together at the computer lab to edit the short videos with MoViE or other editing tool; while preparing for the scene of, for instance, a debate, updating a scenario or shooting from different angles to catch the best shot. In this phase we interviewed small focus groups of (3–5) students according to who was working with whom for story making and telling. Questions asked how group work was organized, what kind of problems came up and how these were dealt with and what were key moments in the digital storytelling experience. Duration of interviews was approximately one hour. To resolve language problems that arise in such international pedagogical research frameworks, the first author, who is Greek, collected observation and interview data from schools in Greece and the second author from Finland. Following transcription, with the contribution of language experts we translated the data in English and held discussions to ensure a mutual understanding of the pedagogical contexts, the meanings conveyed through the interviews and the digital stories, the codes we used for analysis and the themes and categories that emerged.

Contexts of the study

While the majority of Greek students are 6th graders (12-year olds), the 3rd graders (School 2) in this study are faced with short attention span and gender-related issues (e.g., tensions between domineering males and the underrepresented female population of the group) that hinder the learning process in some cases. To resolve such situations, the teachers apply methodologies with more or less structured tasks. In the former students take on specific, alternating roles (e.g., directors, script writers etc.). In other more autonomy oriented learning environments (School 1 and 3) students decide how to distribute roles and reach joint decisions by discussing and voting for or against suggestions and ideas.

The Finnish students of the study are 5th (School 2) and 6th graders (Sc4 and 10) (i.e., 12 and 13 year-olds respectively), choose topics that interest them, write scripts, plan the filming and work independently during and after school hours to make the video stories. One major goal in School 2 and 4 is that students use digital technologies for content, rather than for entertainment only, and learn in a safe networking environment. Another goal (School 2) is to learn how to handle situations where the schedule is tight and invisible technological systems (e.g., bandwidth, device compatibility etc.) impede, even temporarily, the pedagogical process. In Sc10 more technologically expert students teach peers how to use the technologies and MoViE.

Findings and Discussion

Changes in the pedagogical space

Overall, sharing stories and experiences with international peers seems to be an important aspect for all the schools, as students and teachers from all sides admit. Crossing boundaries aiming for an internationally based learning context across the world is equally important. It seems therefore that, if initiative is to be activated and resolve different sorts of unpredictable issues, there needs to exist a source for willingness to do so out there. Getting connected with peers across the world seems to be such a source. Table 1 below presents the themes Finnish and Greek students developed in the stories; how they used the physical learning space; and what medium they used to express themselves.

Table 1. Finnish and Greek student storytelling.

	Finnish students	Greek students
Themes	Recycling, water, respect	Mythology, daily life
Content of learning	Environment and human influence; Well-being; Relationships	The alternation of four seasons; Past and present; Relationships
Learning space in physical proximity	Stories filmed at school, at home, outdoors and in the nature	
Medium of expression	Finnish/ Greek language; subtitles in English; Self presentation videos (mainly) in English; Digital cameras, tablets, smart phones, MoViE platform	

As table 1 shows, students draw themes from the curriculum with focus for instance, on how to improve relationships with peers at school, against animal cruelty, the myth of Persephone that teaches changes in a calendar year and so on. This is a basic requirement for this pedagogical intervention since an overall research goal is to couple storytelling with school everydayness, not to be artificial to it. Although departing from a common formal learning base, the young people find different ways to express themselves and tell the stories with digital cameras and other portable devices. Transferring the story making from formal to informal settings is one unexpected development of this pedagogical experience. Another is that, despite instructed to focus on subject matter content, the students depart to present themselves in an effort to show their international peers who they are and get to know better to one another. These are self-presentation videos of the student, the group and the class and can be seen as expressions of affinity spaces (Gee 2004). By affinity spaces Gee (2004) terms social media environments that encourage the sharing of knowledge or participation in a specific area, and where informal learning is a common outcome. Pedagogical story making and telling seems to be one such affinity space. As Niemi et al. (2014) argue, these are spaces where the youth learn actively by combining fun and engagement with peers and the digital into a meaningful pedagogical dialogue. According to Harju et al. (2014), this is also a space to develop knowledge and skills in collaboration with peers.

Shifting perspectives of self and peers

As table 2 below shows, the analysis progresses from specific (i.e., actual words) to more abstract levels where the 'what' and 'how' of storytelling are described in terms of changes in the pedagogical process. In the next phase main categories and themes emerge.

Table 2. Analysis excerpt: Greek and Finnish student interviews.

Actual words – utterances	Interventions and changes in digital storytelling	Categories and themes
(Gr) S2Sc2-m8: We played, we explored the space and we found a tortoise! It was great!' (Fi) S1Sc4-f13: Well I think it was at the beginning, was it mine or [her] idea that it could be [place] or something that sells soda… But then, I don't know was it her idea when she said that there's a [place] there, so let's go there	We didn't know what it was like; but we took action to explore and get a better insight into alternative urban spaces and our relations with peers	Act to explore

We chose these excerpts to provide evidence of how the analysis developed and, subsequently, the categories that added to our thinking and the evolution process of research questions. As the students transfer the story making into the field, they discover the material dimensions of what it means to set and shoot the digital story. They take on roles and act as professionals while storytelling opens up the opportunity for youth to view themselves as subjects, from a position different from the one of being student. As the male 12-year-old student from Greek school 4 (Student 4, Sc4-m12) states, "I was narrator. I remember we had to shoot an act again and again. It wasn't easy". There, they do not only discover alternatives of the self, they also experience the surrounding environment (e.g., by making 'friends' with a tortoise they find on site). The Finnish students recount similar aspects of the filmmaking event where they meet with opportunities to shift from one role to another and find what suits them best. As the female 13-year-old from School 4 (S2Sc4-f13) explains, "Well, it's like I like to film more and I don't like to perform, and [my friend] likes a lot to be…" "In front of the camera", her partner (S1Sc4-f13) adds.

It seems, however, that there is a shift not only in the way the students perceive the self, but how they look at making and telling stories with peers. One example of such change is when the Greek 12-year-old male (S1Sc4-m12) talks about working in groups, "Initially… I believed I was always right. Now I explain why I support this view without necessarily expecting that everyone will agree". As the 12-year-old recounts his experience, he goes on to add that, "Now I'm interested to know what other people think". By moving from past to present tenses in speech he ultimately expresses not only an interest in how peers think, but also a commitment (Hengel 1994, Ricoeur 1976) regarding the way he envisions his relation with peers from now on. At this point the student takes a critical position toward what used to be known attitude and develops the responsibility necessary for knowing and being through respect and care for peers.

Considering these, it seems that students in this study view being with others under a new light. In this respect, students seem to act for exploration, discovery and learning for content and building relations in both the classroom and the social media environment. Watching the connected peers' digital stories enabled the young people to think twice, suspend the belief of the known and consider alternative ways to shoot the story and ways to act as future citizens. As the Greek 12-year-old female (S1Sc1-f12) explains, "[B]y watching other [students'] videos I could understand things about their habits, their everydayness. Like with the animal testing video, it was so real! And I realized how much they like acting!" In this way the student expresses a shift in perspective in terms of the content of the story and the way the story is presented before she finally expresses a position against animal cruelty.

Likewise, the Finnish students discuss similar aspects of generating ideas by watching stories. As the Finn S3Sc4-m13 explains, "We got to see how they make these videos in other countries... We could compare our own videos with theirs". Again, a Greek 12-year-old female (S3Sc1-f12) seems to align with the view of learning by getting "ideas from peers' videos to make our own". Penttilä et al. (2014) confirm this type of interaction in storytelling in their study on learning science with video stories. In addition, it seems that students discover new ways to tell the story in terms of how scenes are set and how the message is conveyed, by, for example, adding movement rather than still images to the stories. As Kearney (2002) would put it, these young people enter a process of mimesis as discovery.

Conclusions

In this study we used aspects of Ricoeur's (1976, 1992) theory of identity and interpretation in order to discuss the youth initiative for learning with digital storytelling and pedagogical social media. The insight we gained through this experience shows that what keeps the dialogue alive is, in addition to learning by blending literacies and developing new ones, initiative that leads to a shift in the way the young people perceive the self, the peers and the world. In this learning space students develop an embodied rather than exclusively cognitive perspective and display an ability to respond with new interventions and actions. This becomes obvious through mainly three ways of student action. By taking on a role and acting-as, story-making process becomes space for decisions and problem solving. By entering into dialogue with peers, digital storytelling creates a space for acting with respect for others in order to explore and understand deeper and better. In this sense, students do not simply enact a role that has been assigned to them, but learn and grow by changing through own efforts and encourage others to change as well in a community of solidarity and collaboration. It seems therefore that, even in situations of unpredictable outcomes, taking initiative, action and knowing, although not mutually exclusive, are related.

Returning to our initial question whether pedagogies can become transformative through the use of social media and judging from the insights that this study offers, we can respond that, truly, transformative digital pedagogies are feasible. It is also true that we do need initiative on a large scale and such action so that the world transforms into a better place for each and every species that inhabits it with equal opportunities for life, education and prosperity. And yet, we need to start from somewhere. Initiative for interventions and changes do not happen overnight. Neither do they happen in higher education only. If we want the Digital Humanities Pedagogies to have a future it is essential to view education as continuum, in the same way that we need to acknowledge the need to plan flexible curricula and create the space for young people to both envision their social future themselves and have a say in how to shape it.

References

Arendt, Hannah, 1968, *Between Past and Future: Six Exercises in Political Thought*, Revised edition, New York, Viking Press.

Balsamo, Anne, 2000, "Engineering Cultural Studies: The postdisciplinary adventures of mindplayers, fools, and others", in Roddey, Reid and Traweek, Sharon (eds.), *Doing Science + Culture*, New York, Routledge, pp. 259–274.

Bingham, Charles & Sidorkin, Alexander, 2001, "Aesthetics and the Paradox of Educational Relation", in *Journal of Philosophy in Education*, n° 35(1), pp. 21–30.

Boyd, Danah & Marwick, Alice, 2011, "Social Privacy in Networked Publics: Teens' Attitudes, Practices, and Strategies", paper presented at the Oxford Internet Institute Decade in Internet Time Symposium, September 22, 2011.

Freire, Paulo, 2005, *Pedagogy of the Oppressed*, Trans. M. G. Ramos, NY Continuum.

Gee, James Paul, 2004, *Situated Language and Learning: A Critique of Traditional Schooling*, New York, Routledge.

Giroux, Henry, 1994, "Rethinking the Boundaries of Educational Discourse: Modernism, Postmodernism, and Feminism, in Margins in the Classroom: Teaching Literature", in Myrsiades Kostas & Myrsiades, Linda S. (eds), Minneapolis, University of Minnesota Press.

Harju, Vilhelmiina, Viitanen, Kirsi & Vivitsou, Marianna, 2014, "Digital Storytelling in Finnish Schools", in Niemi, Hannele, Multisilta, Jari, Lipponen, Lasse & Vivitsou, Marianna (eds.), *Finnish Innovations & Technologies in Schools A Guide toward New Ecosystems of Learning*, Sense Publishers, pp. 49–56.

Hengel, John van den, 1994, "Paul Ricoeur's Oneself as Another and Practical Theology", in *Theological Studies*, n° 55, pp. 458–480.

Hirsch, Brett D., 2012, "</Parentheses>: Digital Humanities and the Place of Pedagogy", in Hirsch, Brett D. (ed.), *Digital Humanities Pedagogy: Practices, Principles, and Politics*, Cambridge, Open Book Publishers, pp. 3–30.

Hull, Glynda. A., Katz, Mira-Lisa, 2006, "Crafting an Agentive Self: Case Studies of Digital Storytelling", in *Research in the Teaching of English*, n° 41(1): 43–81.

Hull, Glynda A., Kenney, Nora L., Marple, Stacy & Forsman-Schneider Ali, 2006, "Many Versions of the Masculine: An Exploration of Boys' Identity Formation", in *Afterschool Matters Occasional Papers Series # 6*, The Robert Bowne Foundation, Oakland, CA: USA.

Irvine, Colin, 2006, "Moving Beyond the Binaries: A Learning-Centered Approach to Pedagogy", in *Pedagogy* 6(1), pp. 149–153.

Jenkins, Henry, Purushotma, Ravi, Weigel, Margaret, Clinton, Katie & Robison, Alice J., 2009, *Confronting the Challenges of Participatory Culture: Media Education for the 21st Century*, Cambridge, Massachusetts, MIT Press.

Jenkins, Henry, 2012, "From Participatory Culture to Participatory Politics by Way of Participatory Learning", In Proceedings of *Trends in International and Latin American Communication Research* Conference, Santiago, Chile, 18–20 October 2012.

Kearney, Richard, 2002, *On stories*, London, UK, Routledge.

Kill, Melanie, 2012, "Teaching Digital Rhetoric: Wikipedia, Collaboration and the Politics of Free Knowledge", in Hirsch, Brett D. (ed.), *Digital Humanities Pedagogy: Practices, Principles, and Politics*. Cambridge, Open Book Publishers, pp. 389–426.

Kirschenbaum, Matthew G., 2010, "What is Digital Humanities and What's It Doing in English Departments?", in *ADE Bulletin*, n° 150, pp. 55–60.

Kirschenbaum, Matthew G., 2004, "Interface, aesthetics, and usability", in Schreibman, Susan, et al (eds.), *A Companion to Digital Humanities* 2005, pp. 523–542.

Marwick, Alice & Boyd, Danah, 2011, "I Tweet Honestly, I Tweet Passionately: Twitter Users, Context Collapse, and the Imagined Audience", in *New Media and Society*, n° 13, pp. 96–113.

Matusov, Eugene, 2009, *Journey into Dialogic Pedagogy*, New York, Nova Science Publishers, Inc.

Nelson, Mark Evan, Hull, Glynda A., Roche-Smith, Jeeva, 2008, "Challenges of Multimedia Self-Presentation: Taking, and Mistaking, the Show on the Road", in *Written Communication*, n° 25, pp. 415–442.

Niemi, Hannele, Harju, Vilhelmiina, Vivitsou, Marianna, Viitanen, Kirsi, Multisilta, Jari, & Kuokkanen, Anne, 2014, "Digital storytelling for 21st-century Skills in virtual learning environments", in *Creative Education*, n° 5, pp. 657–671. Retrieved from <http://dx.doi.org/10.4236/ce.2014.59078>.

Penttilä, Johanna, Kallunki, Veera and Ojalainen, Johanna, 2014, "Science Through the Camera Lens", in Niemi, Hannele, Multisilta, Jari, Lipponen, Lasse & Vivitsou, Marianna (eds.), *Finnish Innovations & Technologies in Schools A Guide toward New Ecosystems of Learning*, Sense Publishers, pp. 57–66.

Ramsay, Stephen, 2012, "Programming with Humanists: Reflections on Raising an Army of Hacker-Scholars in the Digital Humanities", in Hirsch, Brett D. (ed.), *Digital Humanities Pedagogy: Practices, Principles, and Politics*, Cambridge, Open Book Publishers, pp. 227–239.

Ricoeur, Paul, 1976, *Interpretation Theory: Discourse and the Surplus of Meaning*, Fort Worth, Texas Christian University Press.

Ricoeur, Paul, 1992, *Oneself as Another*, trans. Kathleen Blamey, Chicago, University of Chicago Press.

Saito, Yuriko, 1985, "The Japanese Appreciation of Nature", *British Journal of Aesthetics*, 25: 239–251.

Sartre, Jean-Paul, 1967, *What is Literature?*, Trans. by Bernard Frechtman, London, Methuen.

Simon, Roger, 1988, "Empowerment as a Pedagogy of Possibilty", in *Language Arts* 64: 371.

Short, Harold, 2010, "The Digital Humanities: A Collaborative Discipline", paper presented at the Oxford e-Research Centre, Oxford, May 18, 2010.

Smedt, de, Koenraad, 2002, "Some Reflections on Studies in Humanities", in *Computing, Literary and Linguistic Computing*, 17(1), 92.

Smyth, John, McInerney, Peter & Fish, Tim, 2013, "Blurring the boundaries: from relational learning towards a critical pedagogy of engagement for disengaged disadvantaged young people", in *Pedagogy, Culture & Society*, 21:2, pp. 299–320.

Snell, Patricia, 2012, "Video, Art, and Dialogue: Using the Internet to Create Individual and Social Change", in *The International Journal of Communication and Linguistic Studies*, Vol. 10, Issue 1, pp. 27–38.

Svensson, Patrik, 2010, The Landscape of Digital Humanities, in *Digital Humanities Quarterly,* 4(1). Retrieved from: <http://digitalhumanities.org:8080/dhq/vol/4/1/000080/000080.html>.

Verbeek, Peter-Paul, 2006, "Materializing Morality: Design Ethics and Technological Mediation", in *Science, Technology, and Human Values*, 31(3), pp. 361–380.

Olga Volckaert-Legrier, Josie Bernicot,
Antonine Goumi & Alain Bert-Erboul

Text Messages: Enemy or Ally in the Spelling Learning Process? A Longitudinal Study of 11–12-Year-Old Junior High School Students

Introduction

For the first time in human history, new technologies are providing a great number of people with the opportunity to use written language in the context of interpersonal relationships. Previously, written language had essentially been used in the context of religious practice, administration and laws, the creation and teaching of knowledge, and in literature. Written correspondence between two people had been reserved for a literate minority, while the majority had only very occasionally made use of it in the form of short letters or postcards. By seizing new technologies, today's young adolescents are multiplying their use of written language in text messages, electronic mail, instant messaging, forums for discussion or online homework help, blogs, and social media.

On December 3, 2012, text messaging from mobile phones celebrated 20 years of existence–and of exponential growth, especially among adolescents (12–17 years of age). In France, 86% of adolescents send text messages, claiming they send an average of 435 per week (Bigot & Croute, 2012). Text messages are written using specific spelling forms (textisms) that differ from traditional spelling. Children are using mobile phones at a younger and younger age (Ofcom, 2008), thereby learning traditional writing in school at the same time as they are learning text-message writing. This text-message writing "frightens" parents, teachers, and the media. The hypothesis can be made that text messaging practice has a negative influence on written-language learning at school if text messaging is regarded as an incorrect and degraded form of traditional writing.

This study takes a different point of view by considering text messaging as a new register of written language (Crystal, 2001; Ravid & Tolchinski, 2002), within the framework of pragmatic studies, which attempt to relate linguistic productions to the communicative situation in which they are produced (Austin, 1962; Bernicot, Laval, & Chaminaud, 2007; Clark & Bernicot, 2008; Grice, 1975; Searle, 1969; Verschueren, 1999). A language register is defined as the set of structural language markers appropriate for a given social situation: This implies one does not write a text message to a friend in the same way that one would write a summary of the last geography lesson at school. Studying 11–12-year-old children is especially interesting: They have benefited from a complete–though not yet stabilized–learning process of traditional writing, but with regard to text messaging, they are merely beginning. If text messaging and traditional writing are considered as two registers, the following hypothesis can be made: At the start of text-messaging use, a good traditional spelling level will be linked to the rapid learning of specific forms of the text-messaging register. Once the text-messaging register has taken hold – after several months of text messaging practice – the link between traditional spelling and text-message spelling will no longer exist, as each register will have its own specific application to social situations. The present study tested this hypothesis by using a large corpus of nearly 5,000 text messages produced in daily life situations, and by supplementing existing research primarily carried out with English-language participants.

Relationships between text messages and traditional spelling: previous studies

Previous studies have demonstrated each of the following cases: the absence of correlation between the density of textisms and the traditional spelling level, the absence of effect of text messaging on spelling, a positive correlation indicating that a high density of textisms was linked to a good level of spelling, and the converse. The density of textisms was the essential index used to evaluate the spelling forms used in text messaging: A textism was defined as a change in the spelling form of a word as compared to traditional writing and the density of textisms was equal to the number of changes divided by the total number of words in the text message. In all of these studies, the participants' spelling level was assessed by means of a standardized spelling test.

Absence of correlation between density of textisms and traditional spelling level

Plester, Wood, and Joshi (2009) asked 10–12-year-olds to write text messages on their mobile phones in response to daily life scenarios. These adolescents had approximately 1.5 years of text-messaging experience. The density of textisms produced in these messages was .34 and was not correlated with the spelling score. Plester, Lerkkanen, Linjama, Rasku-Puttonen, and Littleton (2011) analyzed two types of text messages: natural text messages and text messages elicited in experimental conditions following the scenarios of Plester et al. (2009). The participants were Finnish-speaking children between the ages of 9 and 11 who had approximately 2 to 4 years of experience. The average density of textisms was .48 for the natural text messages and .36 for the elicited text messages. Under both conditions, there was no correlation between the density of textisms and the spelling score on the standardized test. The authors confirmed this result by creating two groups: heavy textism users vs. light textism users with regard to the median. The spelling scores of these two groups did not significantly differ.

Absence of effect of text messaging on traditional spelling

Kemp and Bushnell (2011) carried out a study in which 10–12-years-olds were asked to write a message – as dictated by a researcher – on their mobile phone using text-messaging language. The participants had owned a mobile phone for nearly 2 years. There was no difference in the results with regard to the typical manner of writing on a mobile phone (predictive/multi-press/non-texter). The participants who were skilled at multi-press texting and who used more textisms than the other two groups were no different in terms of their traditional spelling scores. Wood, Jackson, Plester, and Wilde (2009) studied the impact of mobile phones on the development of spelling ability in students between the ages of 9 and 10 who had not previously owned a mobile phone. The researchers provided mobile phones to the students for 9 weeks. The students could send text messages on weekends. The text messages sent over the weekend were manually transcribed by the researchers. No significant difference could be seen between the spelling levels of the mobile phone group and the control group before and after the 9-week period. Text messaging practice had therefore not had an effect on the students' spelling skills.

Positive correlation between the density of textisms and the level of traditional spelling

Bouillaud, Chanquoy, and Gombert (2007) compared two dictations (a traditional and a text-message dictation for which the instruction was to write in text-message style) written by students between the ages of 10 and 12. The participants also filled out a questionnaire assessing their practice of new communication technologies. No correlation was found between the questionnaire and either of the two dictations. The text-message dictation was positively correlated with the classic dictation. Plester, Wood, and Bell (2008) asked students between the ages of 10 and 12 to transcribe, by means of a pencil-and-paper exercise, messages from standard English to text-message English. The participants had between 6 months and 1 year 6 months of experience. The average density of textisms was .50. A positive correlation was found between the density of textisms and the spelling test score. Other research has used a longitudinal method to examine natural text messages. In one study, Wood, Jackson, Hart, Plester, and Wilde (2011a) used the experimental paradigm of Wood et al. (2009) to carry out a longitudinal 10-week study of 9- to 10-year-old students who had never owned a mobile phone. The participants could send text messages on weekends and during half-term break. Every Monday, the researchers collated the text messages that the participants had sent. No difference was shown between the scores of the two groups (mobile phone and control group), neither on the pre-test nor on the post-test. In the mobile phone group, the density of textisms remained stable throughout the study. The average density of textisms was positively and significantly correlated with the spelling scores. For the mobile phone group, a high density of textisms was linked to good spelling scores, and conversely. In a different study, Wood, Meachem, Bowyer, Jackson, Tarczynski-Bowles, and Plester

(2011b) spent one academic year recording the density of textisms produced by 8- to 12-years-olds who owned their own mobile phones. The participants had approximately 2 years 6 months of experience. The children were asked to provide a sample of the messages they had sent at the beginning and the end of the school year. The results showed that the average ratio of textisms rose from .33 to .40 between the beginning and the end of the school year. At both the beginning (T1) and the end (T2) of the year, the ratio of textisms was greater for the 11- to 12-year-olds than for the 8- to 9-year-olds. The participants' spelling level remained stable from T1 to T2. The correlations between the density of textisms and the spelling level were positive and significant at T1 and T2. A high density of textisms is therefore linked to good spelling scores, and conversely.

Previous studies have focused on children and young adolescents between the ages of 8 and 12. Diverse methods have been used, from natural data collection to the simulation of writing words in text-message language in a paper-and-pencil situation. The density of textisms has been variable in these studies (from .074 to .53), as has the practice of text messaging (never before, for a few months, for 1 year, 2 years, or 4 years). The longitudinal method necessary to comprehend the acquisition process has not been used, except in the research of Wood et al. (2009), Wood et al. (2011a), and Wood et al. (2011b). It should be noted that in the first two cases, the study period was relatively short (9 and 10 weeks) and in the last case, the study period was one school year–but in that case, the text-messaging samples are only available for the beginning and the end of the year.

The need for a longitudinal study and the particularities of French

In order to go further in understanding the link between textisms and traditional spelling, a method must be used that allows the reliable determination of the density of textisms truly used by the texters: The present study focused on children between the ages of 11 and 12. The methodology of this study enabled text messages to be collected from everyday life situations. The collection was carried out in a longitudinal manner over 12 months with children who had no text-messaging experience prior to the start of the study. This enabled several things to be accomplished: to study the process by which text-message types evolve, to control the duration of text-messaging practice, and to evaluate the link with traditional spelling according to the duration of text-messaging practice. Finally, different types of textisms were examined.

Where the French language is concerned, Bernicot, Volckaert-Legrier, Goumi, and Bert-Erboul (2012a) have proposed two textism categories that are defined by their relationship with traditional writing: a) textisms that are consistent with the traditional written code of grapheme-phoneme correspondence, i.e., the spelling changes do not modify the phonology (pronunciation) of the words and are created with graphic forms existing in traditional writing (e.g., in French, mangez for mangé, kel for quel, and in English, fone for phone, rite for right); b) textisms which break with the traditional code of grapheme-phoneme correspondence, i.e., spelling changes that modify the phonology (pronunciation) of words and/or are created using graphic forms that do not exist in traditional writing (e.g., in French, ri1 for rien, pk for pourquoi, and in English, 2day for today, tmrw for tomorrow). From a cognitive standpoint, the first case is a different application of the same set of rules, while the second case involves the invention of new rules. We hypothesized that, with practice, textisms breaking with the traditional written code would increase more than textisms consistent with the code. Textisms that break with the traditional code are the most typical of the text-messaging register. Therefore, we hypothesized that early on in text-messaging practice, a positive correlation would be shown between the density of textisms breaking with the traditional code and the spelling level. This correlation would disappear after one year of use, once the participants had become familiar with all the characteristics of the SMS register. No correlation between the density of textisms consistent with the traditional code and the spelling level would be shown. The at times contradictory results of previous studies are undoubtedly due, on the one hand, to the type of textism not being taken into account when analyzing the results (except by Plester, Wood, & Bell, 2008), and, on the other hand, to the duration of text-messaging practice not being taken into account (except by Wood et al., 2009 and Wood et al., 2011a).

Collection and coding of the data

Participants

Nineteen sixth and seventh graders having never owned or used a mobile phone participated in the study: 10 girls and 9 boys (mean age = 11.79 years, SD = .59 of a year). The proposal was for the participants to be given access to a mobile phone for one year, free of cost, with the agreement to "donate" at least 20 text messages (written by the students themselves) per month to the research team. All of the participants were from middle-class backgrounds, of legal school age, and native French speakers.

The control group was made up of 30 adolescents having never owned a mobile phone and having never sent text messages. This group without mobile phones was made up of 11 girls and 19 boys (average age = 11.93 years, SD = .59 of a year) who were recruited from the same junior high school and the same classes as the mobile phone group.

Materials

The materials comprised mobile phones which did not have an alphanumeric keyboard and were not flip mobile phones. So as to control and homogenize the text-message sending conditions of all the participants, the T9 key, which enables predictive text and access to a dictionary, was deactivated. All of the text messages in our corpus were written in multipress mode. A software installed on a computer enabled the research team to receive the text messages which were "donated" each month by the participants.

Procedure

Text-message collection

At the beginning of each month, the participants' mobile phones were automatically credited the equivalent of 30 minutes of voice communication or 150 text messages. Once per month, the participants' mobile phones were also credited an unlimited number of text messages to be sent over a period of 5 days. It was during this period that the participants were to send to the research team at least 20 text messages–freely chosen from the text messages that they had sent throughout the month and that they had written themselves. The computer software enabled these text messages to be received onto a computer. This procedure was repeated for 12 months.

Table 1. Examples of text messages from participant no. 13 for month 5. The participant is a boy aged 12 years 1 month.

N° text message	Original text message in French	Text-message transcription in traditional French and English
1	*cc le numéro a rom c 06-XX-XX-XX-XX a tou ta leur ... é cété bien lé heur d'étude ??? é aufaite a l'épiceri il y a dé susette qi color la langue !!! a+*	*coucou le numéro à Rom c'est 06-XX-XX-XX-XX à tout à l'heure... Et c'était bien les heures d'étude ? Et au fait à l'épicerie il y a des sucettes qui colorent la langue ! A plus* hey Rom's number is 06-XX-XX-XX-XX see you later... And how were those hours of studying? And by the way at the store you can get lollipops that color your tongue! Later
2	*si tu regardé plu bele la vie c maren le meurtre*	*si tu regardais Plus Belle La Vie c'est marrant le meurtre* if you watched Plus Belle La Vie the murder is funny
3	*ok mé je sé se qi fo fair je sui pa 1 gamin*	*ok mais je sais ce qu'il faut faire je suis pas un gamin* ok but I know what to do I'm not a kid
4	*pk té pa la???*	*pourquoi t'es pas là ?* why aren't you there?
5	*i a entrenement tou a leur*	*Il y a entraînement tout à l'heure* there's practice later

The participants respected their agreement by providing a total of 4,524 text messages: on average, per participant, 19.84 text messages (SD = 3.02) per month and 238.10 text messages (SD = 36.24) for the year. The data collection took place over the 2009–2010 school year. As an example, Table 1 above presents the text messages (with their original form in

French and their transcribed form in traditional French, and an English translation) of participant no. 13 for month 5.

Evaluation of traditional writing level

At the ninth month of the text-message collection, the two groups took a standardized spelling test from the ANALEC (Analysis of reading ability from 8 years old to adulthood, Inizan, 1998). This test included a classic six-line dictation and a silent dictation (to write beneath pictures the name of what was shown). A total score was calculated, with a minimum score of 0 and a maximum score of 62.

For each trimester, the researchers collected each student's grade in French class (from 0 to 20). It resulted from the combined grades in dictation, grammar, writing, and classical text recitation (poetry or theater). Even though this grade was a non-standardized grade from within the same school and in similar class levels (sixth and seventh grades), it still provided a solid academic evaluation of the student's traditional writing level. For the mobile phone group, each participant's grades in French class at school were also collected for each of the four trimesters during which text messages were collected. For the control group, only the first three trimesters were taken into account, since after this date we were unable to ensure that these students had not acquired or had not used a mobile phone.

Coding of the textisms

Two text-message indexes were considered: type of textisms and density of textisms. A textism was defined as a change in a word's spelling form as compared to traditional writing. For each text message, the density of textisms was equal to the number of changed words divided by the total number of words in the text message.

Two major types of textisms were identified with regard to their consistency or their breaking with the traditional written code (Bernicot, Volckaert-Legrier, Goumi, & Bert-Erboul, 2012a). The different subcategories with examples and equivalences between French and English can be found in Table 2.

In our corpus, a coder-agreement index was measured in the recovery of 10 % of the text messages. The inter-coder agreement value was nearly maximal (99.68 %).

Table 2. Examples of different categories of textisms consistent with or breaking with the traditional code of phoneme-grapheme correspondence.

<table>
<tr><th colspan="5">Textisms consistent with the traditional phoneme-grapheme correspondence code</th></tr>
<tr><td>Categories</td><td>Simplifications</td><td colspan="2">Complexifications</td><td>Substitutions</td></tr>
<tr><td>Examples (and transcriptions into traditional French)</td><td>pe (peux), alor (alors), koi (quoi), ché (chez)</td><td colspan="2">anssien (ancien), est tomber (es tombé), matche (match)</td><td>Chembre (chambre), voie (voix), simpa (sympa)</td></tr>
<tr><td>English equivalent</td><td>hav (have), wen (when), bothA (bother)</td><td colspan="2">Shapping (shaping), leade (lead)</td><td>fone (phone), rite (right)</td></tr>
<tr><th colspan="5">Textims breaking with the traditional phoneme-grapheme correspondence code</th></tr>
<tr><td>Categories</td><td>New phoneme-grapheme correspondences without phonological modification</td><td>Agglutinations</td><td>Phonological modifications</td><td>Words or graphic forms non-existant in traditional French</td></tr>
<tr><td>Examples (and transcriptions into traditional French)</td><td>12C4 (un de ces quatre), CT (c'était), qi (qui), mwa (moi)</td><td>parceque (parce que), ya dlorage (il y a de l'orage)</td><td>nn (non), forene (foraine), aniv (anniversaire)</td><td><3 (cœur), @+ (à plus), Yo (salut)</td></tr>
<tr><td>English equivalent</td><td>2day (today), C U (see you), l8r (later)</td><td>Iloveyou (I love you)</td><td>sn (soon), tmrw (tomorrow), bout (about)</td><td>me bro (my brother), LO (hello), gunna (going to), >– (smiley flower)</td></tr>
</table>

Analysis of the data

The results were analyzed in three stages: textisms, traditional writing, and the link between text messages and traditional spelling. The results of the 12 months were grouped into 4 trimesters, so as to be in line with the progression of the school year.

Textisms

For the variables dealing with the density of textisms, by considering the participants as the random variable, the analysis of the results was carried out with a one-factor ANOVA: text-message use (trimesters 1 to 4). The differences were considered significant at .05.

The dependent variable was the average density of textisms (the total number of textisms divided by the number of words per message) by trimester and by participant. The effect of practice was significant, $F(3,54) = 7.50$, $p < .0003$, $h^2 = .29$, (cf. Figure 1). The density of textisms increased with the trimesters of use. Planned comparisons highlighted a significant difference between trimesters 1 and 2, $F(1,18) = 20.02$, $p < .0003$, and then non-significant differences between trimesters 2 and 3, and trimesters 3 and 4, $F(1,18) < 1$.

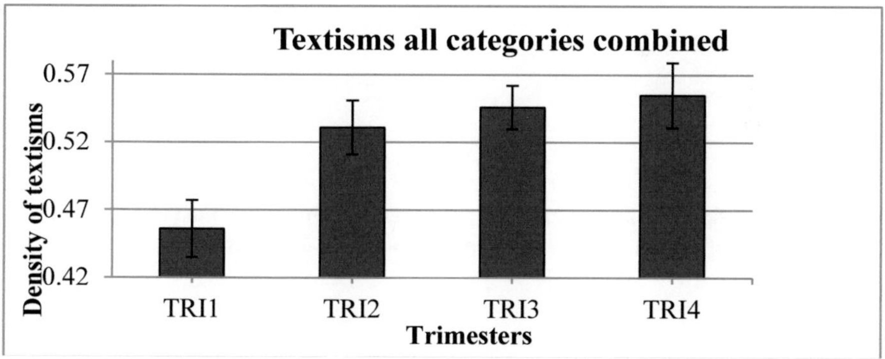

Figure 1: Evolution of density of textisms over one year by trimesters (TRI) of practice. The vertical bars indicate the standard errors.

The average proportion of textisms was .52, signifying that, in a given text message, slightly more than half of the words produced by the participants showed a change as compared to traditional spelling and slightly less than half showed no change. Out of the words containing textisms, half (.26) were consistent with traditional code and the other half (.26) broke with this code.

With regard to the average density of textisms consistent with the traditional code (number of textisms consistent with the code divided by the number of words per message) by trimester and by participant, the effect of practice was not significant, $F(3,54) = 2.15$, $p = .10$, (cf. Figure 2). The participants produced messages containing a density of textisms of .26.

In terms of the average density of textisms breaking with the code by trimester and by participant, the effect of practice was significant, $F(3,54) = 4.21$, $p < .009$, $\eta 2 = .18$ (cf. Figure 2). The density of textisms increased with the trimesters of practice. Planned comparisons revealed a significant difference between trimesters 1 and 2, $F(1,18) = 11.75$, $p < .003$, and non-significant differences between trimesters 2 and 3 and trimesters 3 and 4, $F(1,18) < 1$ (cf. Figure 2).

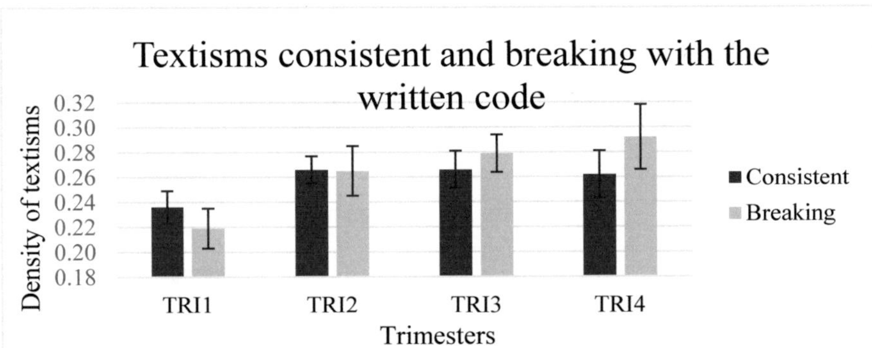

Figure 2: Evolution of textisms consistent and breaking with the written code over one year: by trimesters (TRI). The vertical bars indicate the standard errors.

Traditional writing

We began by comparing the mobile phone group to the control group using two indices: the grades in French at school and the results on the ANALEC standardized spelling test (Inizan, 1998). Table 3 shows the grade in French for the two groups for trimesters 1, 2, and 3. For each trimester, the grades were analyzed using a one-factor ANOVA (group: with mobile phones/without mobile phones). The group effect was never significant (trimester 1, trimester 2, trimester 3: $F < 1$). The two groups took the test at the same moment in time. The score was analyzed using a one-factor ANOVA (group: with mobile phones/without mobile phones). The group effect was not significant ($F < 1$). The mobile phone group scored 48.47/62 on the spelling test, while the control group had a score of 49.03/62.

Table 3. Comparison of traditional writing performance of the mobile phone group and the control group: average school grades in French for trimesters 1, 2, and 3.

	Mobile phone group	Control group
Trimester 1	13.66	14.23
Trimester 2	12.66	12.94
Trimester 3	12.55	12.52

Link between text messages and traditional writing

An analysis was carried out of the correlation (Bravais Pearson's r) between the density of textisms and the ANALEC spelling test scores over the four trimesters of the text-message collection (cf. Table 4).

The correlations that appeared when considering textisms (all categories combined) were the result of two different modes of functioning. Indeed, the separate analysis of the textisms that were consistent and that broke with the code showed that the correlations were not the same for the two types of textisms.

For the textisms consistent with the code, no significant correlation was found between the density of textisms and the ANALEC spelling test score. For the textisms breaking with the code, a positive and significant correlation was found for the first trimester between the density of textisms and the ANALEC spelling test score. The participants that produced numerous textisms also had a good score on the ANALEC test, and conversely.

Table 4. Correlation between the density of textisms and the ANALEC spelling test score.

	ANALEC Stand test Trimester 1	ANALEC Stand test Trimester 2	ANALEC Stand test Trimester 3	ANALEC Stand test Trimester 4
Textisms consistent with the code	0.08	-0.02	0.14	-0.16
Textisms breaking with the code	0.47*	0.32	0.18	-0.21

$^*p<.05$

Conclusion

The data for this study stemmed from a corpus of 4,524 text messages collected over one year from 19 students in sixth and seventh grades (11–12 years of age). With regard to the density of textisms, a result of .52 was obtained, corresponding to the highest results of previous studies, such as that of Plester et al. (2011) in spontaneous production, or that of Kemp and Bushnell (2011) and of Plester, Wood, and Bell (2008) in an experimental situation. When writing a text message, a child of 11 to 12 years of age follows traditional spelling for half of the words, and does not follow it for the other half. It should be noted that this result was obtained in multi-press mode. Kemp and Bushnell (2011) found that, in a situation of elicited text messages, the children writing in multi-press mode had a higher density of textisms (.48) than children writing in predictive mode (.27).

What is the link between textisms and traditional spelling? First of all, it should be emphasized that the comparison between the mobile phone group and the control group showed no difference with regard to French grades in school: The level was identical in trimester 1 and remained identical throughout trimesters 2 and 3. Text-messaging practice, therefore, did not modify the level of French at school. This result is also consistent with an absence of effect of text-messaging practice on the students' spelling level. In accordance with our hypothesis, the textisms breaking with the code increased with practice. Specific to the text-message register, the textisms breaking with the code were learned over the first two trimesters of use, and then their proportion remained stable in trimesters 3 and 4.

Again, in accordance with our hypothesis, the link between the level of traditional spelling and textisms varied with the type of textisms being considered. No correlation was found between the spelling scores and textisms consistent with the code. Positive correlations existed in trimester 1 between the spelling scores and textisms breaking with the code; these correlations disappeared in trimesters 3 and 4.

This study demonstrated the difficulty of answering the question of the link between textisms and traditional spelling in simple terms. The response depends on the type of textisms (consistent with/breaking with the traditional code) and when exactly text messaging was learned (duration of practice). When these elements are not taken into consideration, the present study reproduces the absence of a correlation between textisms and traditional spelling already shown by Plester, Wood, and Joshi (2009) and Plester et al. (2011). When these elements are considered, the correlation can be absent (textisms consistent with the traditional spelling code), positive (textisms breaking with the traditional spelling code and usage-based spelling in the early stages of text-message use), or negative (textisms breaking with the traditional written code and rule-based spelling after one year of text-message use).

What are the academic implications of our results? The first academic implication is that text messages present a new and additional way in which to practice writing. While 52 % of the words in the text messages contained textisms, the other 48 % of the words were written according to traditional rules. It should be emphasized that new technologies are providing students with additional opportunities to practice written language, opportunities that have only recently emerged: e-mail, instant messaging, discussion or online homework help forums, blogs, and social networks such as Facebook and Twitter (Crystal, 2008; Ellison, Steinfield, & Lampe, 2007; Fernandez & Yuldashev, 2011; Herring, Stein, & Virtanen, 2013; Puustinen, Bernicot, & Bert-Erboul, 2011; Puustinen, Volckaert-Legrier, Coquin, & Bernicot, 2009; Savas, 2011; Volckaert-Legrier, Bernicot, & Bert-Erboul, 2009).

The second academic implication is that text messages could be used as an ally for learning at school. This is based on two facts: 1) Adolescents use this form of communication easily and with enthusiasm; 2) No study has shown a negative link between text-messaging and the command of traditional writing. Text messages could therefore be used to support learning at school, as is the case with mobile learning, otherwise known as mLearning (Yengin, Karahoca, Karahoca, & Uzunboylu, 2011). Indeed, UNESCO published a document in 2010 entitled "mLearning: A Platform for Educational Opportunities at the Base of the Pyramid", calling for the development of this type of project.

Acknowledgements

The authors would like to thank all the children and parents who participated in this study, and especially the school that collaborated with the researchers. In addition, the authors would like to thank the students who helped with the collection of the data and Martha Randy for translating the paper.

This research was supported by the French National Research Agency project ANR-08-COMM-011-01 entitled "Adolescents and Computer-Mediated Writing: Prerequisites, Uses, and Learning".

References

Austin, John L., 1962, *How to do things with words*, Cambridge, Massachusetts, Harvard University Press.

Bernicot, Josie, Laval, Virginie & Chaminaud, Stéphanie, 2007, "Nonliteral language forms in children: In what order are they acquired in pragmatics and metapragmatics?", in *Journal of Pragmatics*, n° 39, pp. 2115–2132.

Bernicot, Josie, Volckaert-Legrier, Olga, Goumi, Antonine & Bert-Erboul, Alain, 2012a, "SMS experience and textisms in young adolescents: Presentation of a longitudinally collected corpus", in *Lingvisticae Investigationes*, 35(2), pp. 181–198.

Bernicot, Josie, Volckaert-Legrier, Olga, Goumi, Antonine & Bert-Erboul, Alain, 2012b, "Forms and functions of SMS messages: A study of variations in a corpus written by adolescents", in *Journal of Pragmatics*, 44, pp. 1701–1715.

Bigot, Régis & Croute, Patricia, 2012, *La diffusion des technologies de l'information et de la communication dans la société française*, Paris, CREDOC.

Bouillaud, Céline, Chanquoy, Lucile & Gombert, Jean-Emile, 2007, "Cyberlangage et orthographe : quels effets sur le niveau orthographique des élèves de CM2, 5e et 3e?", in *Bulletin de Psychologie*, 60 (6), pp. 553–565.

Clark, Eve Vivienne & Bernicot, Josie, 2008, "Repetition as ratification: How parents and children place information in common ground", in *Journal of Child Language*, 35(2), pp. 349–372.

Crystal, David, 2001, *Language and the Interne*, Cambridge, Cambridge University Press.

Crystal, David, 2008, *Txtng: The gr8 db*, Oxford, Oxford University Press.

Ellison, Nicole B., Steinfield, Charles & Lampe, Cliff, 2007, "The benefits of Facebook "friends:" Social capital and college students' use of online social network sites", in *Journal of Computer-Mediated Communication*, 12, pp. 1143–1168.

Fernandez, Julieta & Yuldashev, Aziz, 2011, "Variation in the use of general extenders and stuff in instant messaging interactions", in *Journal of Pragmatics*, 43(10), pp. 2610–2626.

Goumi, Antonine, Volckaert-Legrier, Olga, Bernicot, Josie & Bert-Erboul, Alain, 2011, "SMS length and function: A comparative study of 13- to 18-year-old girls and boys", in *European Review of Applied Psychology*, 61(4), pp. 175–184.

Grice, Paul, 1975, "Logic and conversation", in Cole, Peter & Morgan, Jerry (eds.), *Syntax and Semantics*, Vol. 3: Speech Acts, New York, Academic Press.

Herring, Susan, C., Stein, Dieter & Virtanen, Tuija, 2013, *Handbook of pragmatics of computer-mediated communication*, Berlin, Mouton.

Inizan, André, 1998, *Analyse du savoir-lire de 8 ans à l'âge adulte: ANALEC et la dyslexie*, Paris, Editions EAP.

Kemp, Nenagh & Bushnell, Catherine, 2011, "Children's text messaging: abbreviations, input methods and links with literacy", in *Journal of Computer Assisted Learning*, 27, pp. 18–27.

Ofcom, 2008, *Media literacy audit: Report on UK children's media literacy*. Retrieved from <http://stakeholders.ofcom.org.uk/market-data-research/media-literacy/archive/medlitpub/medlitpubrss/ml_childrens08/>.

Plester, Beverly, Lerkkanen, Marja-Kristiina, Linjama, L.J., Rasku-Puttonen, Helena & Littleton, Karen, 2011, "Finnish and UK English pre-teen children's text message language and its relationship with their literacy skills", in *Journal of Computer Assisted Learning*, 27, pp. 37–48.

Plester, Beverly, Wood, Clare & Bell, Victoria, 2008, "Txt msg n school literacy: Does texting and knowledge of text abbreviations adversely affect children's literacy attainment?", in *Literacy*, 42(3), pp. 137–144.

Plester, Beverly, Wood, Clare & Joshi, Puja, 2009, "Exploring the relationship between children's knowledge of text message abbreviations and school literacy outcomes", in *British Journal of Developmental Psychology*, 27, pp. 145–161.

Puustinen, Minna, Bernicot, Josie & Bert-Erboul, Alain, 2011, "Written computer-mediated requests for help by French-speaking students: An analysis of their forms and functions", in *Learning and Instruction*, 21, pp. 281–289.

Puustinen, Minna, Volckaert-Legrier, Olga, Coquin, Danièle & Bernicot, Josie, 2009, "An analysis of students' spontaneous computer-mediated help seeking: A step toward the design of ecologically valid supporting tools", in *Computers and Education*, 53, pp. 1040–1047.

Ravid, Dorit & Tolchinski, Liliana, 2002, "Developing linguistic literacy: a comprehensive model", in *Journal of Child Language*, 29(2), pp. 417–447.

Savas, Perilhan, 2011, "A case study of contextual and individual factors that shape linguistic variation in synchronous text-based computer-mediated communication", in *Journal of Pragmatics*, 43, pp. 298–313.

Searle, John R., 1969, *Speech Acts*, Cambridge, MA, Cambridge University Press.

UNESCO, 2010, *mLearning: A Platform for Educational Opportunities at the Base of the Pyramid*. Retrieved from <http://www.gsma.com/mobilefordevelopment/wpcontent/uploads/2012/04/mlearningaplatformforeducationalopportunitiesatthebaseofthepyramid.pdf>.

Verschueren, Jeff, 1999, *Understanding pragmatics*, London, Arnold.

Volckaert-Legrier, Olga, Bernicot, Josie & Bert-Erboul, Alain, 2009, "Electronic mail, a new written-language register: A study with French-speaking adolescents", in *British Journal of Developmental Psychology*, 27, pp. 163–181.

Wood, Clare, Jackson, Emma, Hart, L., Plester, Beverly & Wilde, Lucy, 2011a, "The effect of text messaging on 9- and 10-year-old children's reading, spelling and phonological processing skills", in *Journal of Computer Assisted Learning*, 27, pp. 28–36.

Wood, Clare, Jackson, Emma, Plester, Beverly & Wilde, Lucy, 2009, "Children's use of mobile phone text messaging and its impact on literacy development in primary school", in *British Educational Communications and Technology Agency* (BECTA).

Wood, Clare, Meachem, Sally, Bowyer, Samantha, Jackson, Emma, Tarczynski-Bowles, M. Luisa & Plester, Beverly, 2011b, "A longitudinal study of children's text messaging and literacy development", in *British Journal of Psychology*, 102(3), pp. 431–442.

Yengin, Ilker, Karahoca, Adem, Karahoca, Dilek & Uzunboylu, Hüseyin, 2011, "Is SMS still alive for education: Analysis of educational potentials of SMS technology?", in *Procedia Computer Science*, 3, pp. 1439–1445.

Sara Minucci & Marta Severo

Strategies for Building Online Identities in Academia. An Exploration of Digital Communication of Researchers in Social Sciences and Humanities

Introduction

In latest years, online social networks have attracted a growing interest among scholars not only as a research field but also as a new space for both professional and personal promotion. The purpose of this essay is to investigate the features and strategies of online communication in the context of academia, in particular among researchers in social sciences and humanities (SSH).

Which are the professional behaviours that researchers in SSH exhibit through their digital profiles? Which tools and strategies do they adopt? Which kind of contents do they diffuse on their homepages? Through empirical analysis, we aim to understand how researchers use the web and social platforms such as Facebook and Twitter as well as professional social networks such as Academia.edu and LinkedIn to promote their research and their personal and professional image. We will pay special attention to the process of building their online identities and social relationships throughout one or more platforms (Casilli, 2010; Cardon, 2008).

After a short state of art of studies about the use of social media in academia, we will cope with our questions through a three-step empirical analysis. Firstly, we analyse the results of an online questionnaire administered to researchers inscribed at the OMERA platform (Observatory of Mobility in Social Sciences in the European Research Area). Secondly, we analyse the factual online image of the researchers that answered the questionnaire on the web. On the basis of the findings of these two phases, we build some hypotheses about identity building in academia that we test through in-depth interviews to researchers. Through such methodology, we expect to give some preliminary answers to our research questions. We will focus especially on the three facets of the use of the web in academia: (1) how researchers use the web and social platforms to promote *their personal and professional image*; (2) how researchers use the web and social platforms to *build their online identities*; and (3) which relationship does exist between the image that the researcher intends to build and *the image that factually appears on the web*.

Social media in academia

The Internet constitutes an important way for self-presentation (Cardon, 2008). It is also highly efficient for networking and expanding the number of connected people. The importance of social relationships in academia was already noticed by Gersick et al. (2000) that through empirical analysis verified the impact of social ties in academic professional life and notably for the career development. Actually the web, and online social networks in particular, represent great tools for scholars because they help them carrying on a series of actions that are central in their work such as gathering data and literature, networking, spreading ideas and concepts, overcoming the borders among different disciplines and different audiences. Furthermore, social networks, especially those designed specifically for academics, such as Academia.edu, can be very efficient in engaging with scholars working in the same discipline. Thus, recent literature has showed a growing interest in understanding the different uses of the web among academics (Katz, 2010) and in the use of social media by scholars (i.e. Veletsianos, 2012 on the use of Twitter). Recently Weller and Strohmaier (2014) have studied the impact of social media in academia yet they focused on social media as source of data for research more than on their impact on the professional life and online identity of the researcher. There is also a branch of literature focusing on the possible influences of social networks and blogs in academic teaching (Pierce et al. 2011; Ranieri, 2014).

In this context, our work aims at changing the point of view from the tools to the actor. In this direction, Celik et al. (2014) worked specifically on the academics' perception of social media. They investigated the effect of social media on the researchers' motivation. With a similar extent, Jeng et al. (2014) carried out a case study on Mendeley users and they found out that "users of Mendeley seem to mainly concentrate on the utilities directly related to their research work, while mostly ignoring its social features, such as 'friend making'" (2014: 15). On the basis of these first results, this study is meant to investigate more deeply the relationship between scholars in SSH and the opportunities the web offers. This is why the starting point for our analysis is researchers' habits on the web.

Methodology

Data for this analysis have been collected thanks to OMERA Social Sciences, a project aiming at exploring the impact of the European Research Area (ERA) policies on SSH by focusing on the practices of researchers and grant-makers (such as foundations and research institutes) in the emerging European research market. The project is developed through a website <www.omerasocialscience.org> gathering funding opportunities in SSH research from more than 90 funding institutions. The website is open for consultation to registered researchers from all over the world.

We developed our empirical analysis in three phases. Firstly, we administered an online questionnaire to OMERA subscribers using its mailing list. Through questionnaire's data, we will describe which is the relation between researchers and digital tools both for professional and personal communication, and which strategies researchers adopt for building their online identity.

Secondly, we analysed the factual online image of the respondents. In order to do this, we carried out two empirical investigations: (i) a pilot ethnographic exploration of personal web pages that researchers mentioned in the questionnaire. This technique allowed us to analyse the different strategies adopted by respondents in their websites. (ii) Then we verified the online identity of the researcher as it appears on Google results. We especially focused on the integration between personal and professional identity.

Thirdly, on the basis of the comparison between respondents' declarations and their factual image on the web, we make the hypothesis that the different behaviours of researchers for building their online identity can be reduced to three main strategies. In this final phase we tested such hypothesis through in-depth interviews to a sample of users that answered to the OMERA questionnaire.

Results

The OMERA questionnaire: the perspective of researchers

The questionnaire (constituted by 20 multiple choice questions and 4 open questions) was sent at the beginning of March 2013 and remained online until the end of April 2013. At that time, users registered on <www.omerasocialsciences.org> were 434. We obtained 43 answers (9,9 %): 27 from women, 16 from men. The most part of researchers who answered is between 30 and 40 years old (58 %).

Respondents' nationality is mostly Italian (51 %, even if only 41 % actually works in Italy), followed by Polish and Portuguese (9 %), and by Spanish and American (4 %). Other nationalities represented are: Argentina, Belgium, Benin, Brazil, Canada, Czech Republic, France, Greece, Morocco.

The majority of respondents are at the beginning of their career: they are actually Research Fellows (58 %) or PhD Candidates (30 %), while only 2 % are Full Professors. Furthermore, 33 % works in a university, 9 % in a foundation, 5 % in a research institute, 7 % both in a university and in a research institute, while 2 % at the moment of their answers had no affiliation at all.

The questionnaire revealed that 77 % of the respondents consider the internet really important in researchers' work and they broadly use it as a tool to manage a professional digital profile, since the 84 % thinks that an online personal profile is useful for her/his work. It is important to underline that 44 % of the scholars think researchers are pushed to have a digital profile by their own working environment and just the 28 % does not feel this pressure to open a public personal account on the web.

Taking into account these premises, it does not surprise the fact that they appear to be present on the most popular social networks (multiple answers were possible) [1]: 67 % has a profile on Facebook, 42 % on Twitter, while only 7 % on Pinterest. Respondents also declare having a page on professional social networks, in particular Academia.edu (67 %) and LinkedIn (65 %). Furthermore, researchers seem to be interested in managing their profiles on social networks: 28 % check them everyday and 19 % five days a week, 21 % once a week. Only 12 % check them only once or twice a month.

When considering other forms of online presence, 56 % of the scholars declares to have a personal page on his/her working institution official website. It is interesting that almost all those who declared to have an institutional page work in university. Interestingly, the majority of those having a personal page on their working institution's official website (44 %) consider it part of their job, while 5 % declares that it constitutes an extra part of researcher job but it is useful for his/her work. Among our respondents, only the 20 % declares that his/her institutional page contains direct links to other pages (of those, 40 % is linked to Academia.edu, 10 % to LinkedIn, 10 % to Twitter, 10 % to the scholar' personal blog, 10 % to other research institutions pages, 10 % to other researchers blogs, 10 % to other researchers profiles).

On the contrary, 83,3 % of the scholars admitting to have a blog (14 % of total respondents) not only declares that it is meant to address both to academics and non academics, but admit also that their blogs contain links to other pages: 27 % is linked to Twitter, 18 % to Facebook, 18 % to other researchers blogs, 18 % to other researchers profiles, 9 % to LinkedIn, 9 % to his/her institutional profile on his/her working institution official site). The scarce result obtained by blogs can be explained by the strong time consumption necessary to their management.

Our respondents appear to have a much more 1.0 or 'static' approach to their institutional web profile. On the contrary, they seem much more interested in networking and in taking advantage of the possibilities offered by web 2.0 when managing their personal blogs or their profiles on social networks. Furthermore, scholars having a profile on one or more social networks are significantly more that those having an institutional page. This could be explained by the fact that the most part of our respondents is at the early stage of the career, a period that clearly leads them to a

---

1 Researchgate was not proposed as an answer because, according to official statistics of 2013, this platform counted 2 million users compared to the 10 millions of Academia.edu.

high level of mobility across different institutions and also across different countries, so they might want to invest more in a personal profile than in an institutional one.

Screening the Web: The factual image of researchers

As a second phase, we analysed the factual online image of the respondents. We could perform this analysis on 18 interviewees that accepted to renounce to their anonymity.

We organised our investigation in two steps. Firstly, we analysed the professional identity of the researcher by exploring the personal web pages, accounts and blogs that he/she mentioned in the questionnaire. This phase is expected to help us to recognize the different strategies that researchers are implementing in order to build their online professional image. We verified: the amount of published content, the updating periodicity, the type of content (publications, news or critics/analysis), and the variety of content among the different accounts.

We visited globally 58 accounts/web pages/blogs. We could not visit few pages: some because there were offline; and others were impossible to find because the researcher did not provide the exact address and it did not appear in the Google's results neither by searching directly on the platform especially for LinkedIn and Facebook accounts. We may suppose that the researcher abandoned or deleted such web pages in the months following the questionnaire or that he/she selected strong privacy settings to avoid to be found by search engines.

We can see that, as in the general sample, most of researchers have chosen Acamedia (72 %) and LinkedIn (67 %) as the main tools for building their identity. Then, half of respondents have also a Twitter account (a little more than the general sample) and a Facebook account (a little less than the general sample). 4 people (22 %) have a blog and 11 (61 %) have a personal page on the website of their institution.

We identified some general trends. Generally, the researcher does not update all these proxies of his/her image at the same time and with the same content. Conversely, he/she fills the different platforms' accounts with varied content and periodicity. We can make the hypothesis that this selection is effect of the lack of time that the researcher can dedicate to update his/her multiple accounts, so that he/she had to prioritize one or two platforms.

For who has chosen Academia.edu, it becomes the main tool for gain visibility and improving the professional contacts' network. As summarized in Table 1, these profiles generally include several publications and obtain a high number of visits. Only 5 people (all over 30 years old) do not have an Academia.edu account. Actually, two among them opened an account in the months following the questionnaire. Most of people using this platform updated the profile during the last month and published on it most of his/her publications (up to 99). We can notice that researchers do not have big network of followers (less than 100, except for 2) and the number of people they follow is almost equivalent of the number of people that follow them. We can highlight the high number of documents' and profile's views that several account reach, regardless with their professional status. Taking into account this big variety of views' amount (from 59 to 3418) in the sample, we can make the hypothesis that the level of visibility of each account is related to the promotion strategy that the researcher put into practice (and not simply to the amount of visits coming from Google). The researcher has several options to promote the Academia.edu account: he/she could send links to his/her publications through the Academia.edu platform or can include the URL address of his/her profile in the curriculum vitae or in the business card.

Table 1. Activity on Academia.edu.

Age	status	documents	views	documents views	followers	following	last update
25–30	PhD	13	893	519	146	78	11 days ago
25–30	PhD	19	710	473	83	78	8 days ago
35–40	Post-Doc	No account					
30–35	Post-Doc	15	525	167	29	24	4 days ago
25–30	PhD	0	116	0	44	54	8 days ago
25–30	Post-grad	9	294	168	96	86	5 days ago

Age	status	documents	views	documents views	followers	following	last update
35–40	Faculty	99	2211	3811	198	124	21 days ago
35–40	PhD	4	214	58	22	30	25 days ago
30–35	PhD	No account					
40–45	Post-grad	No account					
> 45	Post-Doc	19	59	891	12	2	8 months ago
30–35	Faculty	No account					
25–30	PhD	5	1029	113	88	87	17 days
30–35	Post-Doc	No account					
35–40	PhD	11	1218	314	32	40	30 days
30–35	Faculty	6	126	54	11	4	9 months
30–35	Faculty	14	1120	519	67	56	20 days
30–35	Post-Doc	46	3418	2968	243	413	4 days

As regards LinkedIn, the use of this tool is really various. Generally researchers create an account and they fill it with various professional content, then the maintenance is managed in a very different way. Few accounts (3) have a very rich page wand a big network (up to 500 contacts). Yet, half of the analysed profiles have few connections and publish rarely some up-to-dated content. What is really interesting to note is that four people keep their account private, accessible only for their personal contacts. It means that LinkedIn is not considered a tool for improving their visibility and to enlarge their network, yet just a platform where they need "to be" in order to maintain contacts they met in other kind of occasions (colleagues, school mates, rarely people met during conferences). Finally, we can observe that several accounts (at least 3, but we do not know for the private accounts) on LinkedIn are not updated at all: they are created and rapidly forgotten.

Then, if we consider the activity on Facebook, it is really varied. First of all, we need to precise that we could visit only 5 accounts because an account was private, accessible only for friends, and other four were not possible to find. It means that they are used only for personal purposes, so they were not so interesting for our analysis. Yet, we found few researchers (4) that keep their Facebook accounts completely public. In these cases, Facebook walls generally combine personal and professional updates. People doing this are clearly oriented to gain more visibility even mixing the personal and professional image.

As regards Twitter, in the questionnaire 9 people declare to have an account. Yet, one was offline, one was open and then abandoned, another one had no update since one year. The other 6 are really interesting (Table 2). Twitter is used as a professional tool for diffusing news related to the research sector of the researcher. In 2 cases, professional and personal contents are mixed, but we can notice that generally these accounts are employed for following and diffusing news related to the research sector. The choice of opening a Twitter account can be considered a good indicator of the willingness of the researcher of promoting his/her professional image on the Web. Considering that most of accounts have more people following than followers, we can presume that the impact of this strategy of visibility is limited.

Table 2. Activity on Twitter.com.

Age	status	tweets	following	followers	type of content	last update
25–30	PhD	127	21	52	pro	4 days
25–30	PhD	449	406	91	pers+pro	10 days
35–40	PostDoc	918	370	185	pro	5 days
30–35	PostDoc	542	308	118	per+pro	5 days
25–30	Postgrad	0	10	0		
35–40	Factuly	141	400	107	pro	1 day
40–45	Postgrad	160	485	131	pro (retweet)	1 day
> 45	PostDoc	not found				
30–35	Faculty	30	81	63	pro	1 year

As a second phase, we tried to measure the actual visibility of the researcher's professional identity by observing as it appears in Google. By googling their names (in the national version of Google of the country where the researcher works), we verified: (i) how many of the first ten results concern actually the interviewee; (ii) if the results were related to her/his professional or personal identity; and finally (iii) if the professional results included the social network accounts mentioned in the questionnaire. Results are summarized in Table 3.

Generally, we found that the respondents have a good visibility on Google: almost all ten first results concerned the searched person. In two cases, it was necessary to complete the query with the name of the institution in order to disambiguate with homonyms.

For all respondents, almost all results concern their professional life. We can observe few cases where a Facebook (or ones a Twitter) account appears in the results, yet it was not mentioned in the questionnaire. So we can make the hypothesis that the respondent considered this account not useful for his/her professional identity. In other cases, the account is of a homonym.

Table 3. Visibility on Google.

Age	status	Mentioned profiles	Found profiles	First result	Other results
25–30	PhD	Facebook, Twitter, Academia, LinkedIn, blog	Facebook, Twitter, Academia, LinkedIn, blog	Linkedin	Researchgate
25–30	PhD	Twitter, Academia, LinkedIn, institut. page	Academia, LinkedIn, institut. page	Academia	Facebook
35–40	Post-Doc	Twitter, LinkedIn, blog, institut. page	Twitter, LinkedIn, institut. page	institut. page	Academia, Facebook (abandoned blog)
30–35	Post-Doc	Twitter, Academia, institut. page	Academia, institut. page	institut. page	
25–30	PhD	Facebook, Academia, LinkedIn, institut. page	Facebook, Academia, LinkedIn, institut. page	Academia	
25–30	Post-grad	Facebook, Twitter, Academia, LinkedIn	Facebook Academia, LinkedIn	Academia	Other networks such Scribd, Photoblur

Age	status	Mentioned profiles	Found profiles	First result	Other results
35–40	Lecturer	Facebook, Twitter, Academia, LinkedIn, blog, institut. page	Academia, LinkedIn, institut. page	institut. page	Google site, Amazon
35–40	PhD	Academia, LinkedIn	Academia	Academia	
30–35	PhD	blog, institut. page	blog, institut. page	institut. page	Several events and publications
40–45	Post-grad	Facebook, Twitter, LinkedIn	Facebook, Twitter	Media	
> 45	Post-Doc	Facebook, Twitter, Academia, LinkedIn, institut. page	Facebook, Twitter, Academia, LinkedIn, institut. page	institut. page	
30–35	Research Fellow	Facebook, Twitter	Twitter	Twitter	Academia, Google+, LinkedIn
25–30	PhD	Facebook, Academia, Institut. page LinkedIn,	Academia	Academia	
30–35	Post-Doc	Facebook, LinkedIn	Facebook	Facebook	Amazon
35–40	PhD	Academia, Institut. page	Academia, Institut. page	Academia	Vimeo, Google+
30–35	Lecturer	Academia, Institut. page	Academia, Institut. page	institut. page	
30–35	Lecturer	Academia, Institut. page	Academia, Institut. page	institut. page	
30–35	Post-Doc	Academia, LinkedIn, Institut. page	Academia, LinkedIn, Institut. page	institut. page	Google scholar

What is interesting to note is that professional results rarely include all the platforms created by the researcher. On the one hand, more the researcher is active more is probable that there exists several web pages that include his/her name, such as pages concerning events where he/she took part, or pages that present his/her publications, or also media participations (talk show, newspapers articles), or the pages of other professional networks. For several respondents, we found also their institutional web page in the previous institutions where they worked. Some of these pages can be part of visible websites such as Amazon that clearly appear upper in the Google results than the Twitter or LinkedIn accounts of the researcher. On the other hand, we can notice that, even if the respondent has a lower notoriety on the web and there are few pages mentioning him, it does not imply the all the accounts he/she has created for improving his/her visibility will appear on the first results of Google. If these accounts are poor in content and rarely updated, it is like if they do not exist. In general, we found rarely LinkedIn and Twitter accounts.

Finally, we can observe that generally from PhD and postgraduate students the first result is one of their account (mostly on Academia.edu), while for research fellows the first result is the official page dedicated to the researcher in the website of his/her institution. We can highlight how search engines do not easily award the efforts of the researcher in building his/her professional image. Even if the respondent declares to have several professional accounts and to update them, it does not mean that those accounts have a clear impact on his/her professional image.

The strategies for building online identity in academia

The first two empirical analyses helped us to highlight two complementary aspects of building an online identity in academia. On the one hand, the relevance of having an online presence emerged in the questionnaire, where almost half of the scholars admitted to feel pushed by their working environment to manage a digital profile. This is a crucial aspect because it shows the importance gained by the web also for academic marketing. On the other hand, through the second analysis, we observed that, if our respondents have generally a good visibility when googling their names, factual online identity can be really different among respondents. If for someone accounts on professional social networks are at the top of the Google results, for other they are almost invisible and other kinds of pages appear as first results of the search engine.

As a third step of the analysis, through the comparison of the respondents' declarations in the questionnaire and the factual image of respondents as emerged by surfing the web, we can identify three main strategies adopted by our respondents for building their online identity.

(1) The first strategy corresponds to researchers that feel the *obligation* to be online generally because of their working environment, yet they do not update their profiles. We will call them 'digital convicts'.
(1) The second strategy is developed by researchers exclusively oriented to improve their professional image. They can have different status, from PhD to faculty member. They focus their strategy on Academia.edu, yet they can use also LinkedIn and sometimes Twitter. We call them 'digital focused'.
(2) The third strategy corresponds to researchers that are oriented to gain visibility in general. They use all platforms, mixing personal and professional elements. We will call them 'digital performers'.

In order to verify the hypothesis of the existence of these three strategies, we carried out 9 in-depth interviews with some of the respondents that participated to the previous two analyses. The interviewed presented a varied composition in order to represent the characteristics of the general sample: we have 3 PhD candidates, 4 post-doc and 2 researchers. 5 of them work in Italy, one in Australia, one in Spain, one in Portugal and one in Ireland. 5 of our respondents lay in the age class 30–35, 3 of them are between 25 and 30 years old (so they are digital natives (Prensky, 2001) and only one is in the age class 40–45. Interviews were developed in September 2014 and their average duration was 25 minutes. We proposed 7 questions in order to investigate:

(1) If respondents feel a pressure to have an online identity or/and they feel the usefulness of it;
(2) If respondents have a *strategy* to gain visibility and if they verify their online identity;
(3) If their strategy concerns just their *professional identity* or their general image.

As regards the first research question, we found that the majority of our respondents seem to feel a sort of 'social pressure' in being online coming mainly from the academic environment. This kind of pressure concerns in particular their personal pages on their institutions web pages that are used by institutions for evaluation purposes in all the countries in which our respondents work. In some cases they explained us that their working contract requires to keep their institutional personal pages always updated. In general, lot of them appear worried about the relevance that the online presence has gained and also about how this is 'changing the rules' in academia: "Let's say that sometimes when I receive email from some colleagues that at the bottom have 'last article xy' or colleagues that have a more systematic use of […] tools of self-promotion […] it makes me wonder about the structure of the academic editorial market and about the transformations of the profession rules. […] Sometimes I ask myself if it is something I want to go along with." [Researcher A, Italy].

Also, the relevance of the role of the online profile seems to be linked to the subject of their research, there are themes that are still linked to other networks, for example when working on very local phenomena: "In my specific case [digital profiles] don't help me that much because I'm working on very local issues and then networks work in other way" [Lecturer, Italy]. While other broader 'subject communities', i.e. Middle East, can keep a real advantage of the possibilities of sharing materials and informations via academic digital profiles: "I remember that two years ago we organised a panel at the SISP annual conference, I put the call for paper there [on Academia.edu] and it worked: we received proposals that we wouldn't have received in other ways" [PostDoc, Ireland].

Moreover, all our respondents agree about the importance that being on the Internet with professional profiles has and they are also aware of the advantage they can take through the visibility that online profiles give to their work. They do not seem all convinced that a strong online presence does not coincide with an equivalent academic reputation, that is one of the reasons why, as we will see later, most of them do not follow a strategy in organising their digital profiles. Namely, some of them appears convinced that citations still come from articles published on traditional academic journals, while from the presence of their work on academic social networks can come new collaborations and new works: "Through ResearchGate I spread the most part of my publications so other researchers know that I work on themes they didn't know about and sometimes from these shared works new four handed works have born." [PostDoc, Spain]. While others suspect that digital profiles alone can give great credibility to researchers: "Credibility is always more linked to the mere online presence, not necessarily on the quality of the work" [Ph.D., Australia]. Also these different approaches highlight the fact that the profession of the researcher is changing and that digital profiles and practices are in a way helping to booster this process.

Concerning the second research question, we found that our respondents demonstrate not to have an established strategy in managing their online presence even if they recognise the role of the internet in giving them visibility for their works: "The fact of being able to give visibility to things that can generate new visibility is surely the criterion I use the most" [PostDoc, Ireland]. This attitude is mostly justified by the fact that keeping update their digital profiles is considered a highly time consuming activity and this is also the reason why not only they rarely demonstrate to have a plan for boosting their online visibility but also why even when they have pages on many different social networks they do not seem to target their contents or to use popular social networks (i.e. Twitter) to promote their publications.

Furthermore, our respondents do not seem to have a strategy to promote their online profiles both online and offline: generally, they do not give the link to their accounts during public presentations and they do not have links to their social profiles at the bottom of their emails. Even if, interestingly enough, some of our respondents reported to include links to their online profiles when writing emails to someone they do not know personally, to present their work and their position in academy: "The only thing I do [to promote my digital profiles] is to put the link to my page when I have to introduce myself to someone to ask something or to propose something" [PostDoc, Italy]. In this way, digital profiles seem to be used by our respondents as business cards while it does not happen the contrary: our respondents' business cards do not carry any reference to digital profiles. "It's useful [to have digital profiles] because I don't bring my business card around so I can always give the reference to my digital profile at conferences or when I meet colleagues" [PostDoc, Portugal].

It is interesting to underline that, even if they do not have a strategy to promote their digital profiles, all our respondents admit to google his/her name to see how visible they and their works are and they are all aware of the relevance their digital profiles have to offer visibility to their works "Let's say that I try to use it to my advantage. […] In this sense it should be positive. When there is a new data aggregator I try to log me in, to use it." [Ph.D., Italy]. This 'googling fever' is particularly linked to the evaluation process to which everyone of them is subjected to. Also the social networks themselves seem to push researchers in paying attention to the results their digital profiles obtain, as some of our respondents pointed out: "I feel a pressure coming [...] from social networks. As the fact that they send you messages, mails, suggesting you to follow other people and to check who published what. [...] I check [my profile on] Academia to understand who is searching for myself and my articles or who finds my name, which articles are read the most" [PostDoc, Italy].

For what concerns the third research question, we found that the only strategy that almost all our respondents seem to adopt is to keep separate their professional and personal digital profiles: "My profiles are about my profession, only my Facebook profile is personal [...] and my friends there are almost exclusively friends" [Ph.D., Italy]. When talking about their use of social networks, it comes to light how our respondents tend to use them almost exclusively for professional purposes: "If I am in a particularly nice place I can upload a picture and say 'I'm in a fantastic place'. For me this is the maximum level of 'personal' that I share online. For other things [not professional] I use other channels" [PostDoc, Portugal]. Facebook is the only platform on which some of them seem to feel free to mix together their personal and personal identity. Only one of them declared that she has two Facebook profiles, an 'official' one and a more 'friendly' and free one with a nickname.

The interviews we developed helped us to verify our hypothesis and to deepen the understanding of our respondents use of their digital profiles. In this way, we can say that the majority of researchers we met can be classified as 'digital focused' since they tend to have only a professional identity online. Interestingly, both among respondents that can be labelled 'digital focused' and those we called 'digital performers', there are not age-related differences since all age classes are presented.

For what concerns the 'digital convicts', only one of our respondents can be classified as such even if it is important to underline that most cases a kind of pressure is perceived by researchers in being online: it can come from their working environment but it can also come and be fostered by social networks rules and promotion strategies. The three profiles we identified can thus represent in a way three different ways of reacting to the changes that are going on in researcher' profession also thanks to the prominent role of the internet in academia.

Conclusions

This essay is meant to study the digital communication of researchers in social sciences. Literature on this topic is quite recent and sparse. Even if several scholars have recognised the impact of social media in academia, and notably on students' attitudes and behaviours, few studies have tried to measured the impact on academics' activities. These few studies generally focused on the effects of digital communication on specific aspects such as the career development or the motivation for using such platform. Our research tries to adopt a more general approach and to investigate more in general the uses of Web 2.0 platforms by researchers. The aim is to define the strategies that usually scholars adopt for building their online identity.

In order to analyse such a complex object, we built a case study organised in three phases. On a first phase, we proposed a questionnaire to a large sample of researchers related to the OMERA platform. The relevance of having an online presence clearly emerged in the questionnaire, where almost half of the scholars admitted to feel pushed by their working environment to manage a digital profile. Than on a smaller sample, we observed their uses of different digital platforms and their online visibility as it appears through Google. What emerged is that respondents have a good visibility when googling their names, yet the type of results and the order of the different profiles is quite various. Generally, search engines do not easily award the efforts of the researcher in building his/her professional image through the different profiles on official websites and social platforms. Yet, search engines seem more sensitive to the fact that researcher is mentioned on big websites such as Amazon.

On the basis of these first results, we developed some hypotheses about the strategies adopted by researchers for building their online identities. We were particularly interested in verifying if researchers feel just the obligation to have an online identity or if they are sincerely motivated. And in this second case, we tried to verify if they have an aware strategy and if such strategy manage the distinction between personal and professional profile. As a result, we found three profiles: the digital convicts, the digital focused and the digital performers. We can noticed that is the digital focused was the more frequent profile in our sample, we found only a 'digital performer'.

As a conclusion, it is important to underline that those are only preliminary results and further research is still necessary on this young field. Yet, we consider these first hypotheses quite promising. Future analyses can be carried out for distinguishing strategies through time (different phases of the carrier) and through space (different country or type of organisation).

References

Casilli, Antonio, 2010, *Les liaisons numériques*, Paris, Edition du seuil.

Cardon, Dominique, 2008, " Le design de la visibilité : un essai de cartographie du web 2.0 ", in *Réseaux*, n° 152, Paris, Lavoisier, pp. 93–137.

Celik, I., Akilli, G. K., & Onuk, T. C., 2014, "Social Media for Academics: Motivation Killer or Booster?", in *Advanced Learning Technologies* (ICALT), pp. 480–482.

Davis, Charles H., Deil-Amen, Regina, Rios-Aguilar, Cecilia, & Gonzalez Canche, Manuel S., 2012, *Social Media in Higher Education: A literature review and research directions*. Retrieved from <http://works.bepress.com/hfdavis/2/>.

Gersick, Connie J., Dutton, Jane E., & Bartunek, Jean M. ,2000, "Learning from academia: The importance of relationships in professional life", in *Academy of Management Journal*, 43(6), pp. 1026–1044.

Jeng, Wei, He, Daqing, & Jiang, Jiepu, 2014, *User Participation in an Academic Social Networking Service: A Survey of Open Group Users on Mendeley*, in arXiv preprint arXiv. 1401.6495.

Katz, Richard, 2010, "Scholars, scholarship, and the scholarly enterprise in the digital age", in *Educase Review*, 45, pp. 44–56. Retrieved from <http://www.educause.edu/ero/article/scholars-scholarship-and-scholarly-enterprise-digital-age>.

Pearce, Nick, Weller, Martin, Scanlon, Eileen, and Kinsley, Sam, 2010, "Digital scholarship considered: how new technologies could transform academic work", in *Education*, 16 (1), pp. 33–44.

Prensky, Marc, 2001, "Digital natives, digital immigrants part 1", in *On the horizon* 9.5, pp. 1–6.

Ranieri, Maria, 2014, "Le competenze digitali dei giovani ricercatori. Quadro teorico, modelli di analisi, proposte formative", in *Pedagogia Oggi*, n° 1, pp. 180–198.

Veletsianos, George, 2012, "Higher education scholars' participation and practices on Twitter", in *Journal of Computer Assisted Learning*, 28 (4), pp. 336–349.

Weller, Katrin & Strohmaier, Markus, 2014, "Social media in academia: How the Social Web is changing academic practice and becoming a new source for research data", in *it-Information Technology*, 56(5), pp. 203–206.

Danielle Masterton

Identity in the Era of Big Data: An Embodied Approach

Introduction

This paper investigates the implication of identity in big data. Whilst the co-imbrication of bodies and technologies is explored in critical and popular discourse, in the context of big data this interconnection is seldom addressed. A reconsideration of the big data revolution in relation to ontology is essential because we have witnessed a paradigmatic change in how we communicate and interact with the world. Although the majority of data collection occurs in machine-to-machine transactions (WEF, 2013: 7), big data most often concerns human subjects. Indeed, according to statistics produced by the International Data Corporation, 'individuals' direct or indirect actions generated about 70 per cent of the digital data created in 2010' (in WEF, 2011: 9). The increasing enmeshing of personal data in the constitution of big data urges a re-evaluation of the way in which identity is implicated. What is more, the big data revolution not only comprises a shift in the complexity and capacity of analytics but also the incorporation of this data into the global economy.

Views about data emerge in two distinct camps. On the one hand, data is conceptualized as personal, intimately pertaining to the individual from whom it originated, resulting in a conceptualization of the data subject as an essential co-creator in the socio-economic value at stake. This view is exemplified by the European Commission approach to personal data, envisioning data as constitutive part of the self (EUJustice, 2012). On the other hand, industry evaluations of data envisage a new economic asset. The World Economic Forum pan-national initiative "Unlocking the Value of Personal Data" typifies the declaration of data as an asset (2013). The implication of individuals combined with the immense potential of value creation for industry and government creates conflicting values that demand negotiation; the challenge, as Tennenhouse asserts is "to protect individuals without destroying data's socio-economic potential" (2013). Big data is caught between the opposing pulls of industry profit and the protection of individuals and identities. The juxtaposition of these contradictory trends creates a space in which to question how we can or should value data, and by extension identity, in the era of big data.

Personal data: A new asset

The agenda set out by the World Economic Forum's multi-year initiative "Rethinking Personal Data" proposes big data as a new economic asset; "beyond its sheer volume, data is becoming a new type of raw material that's on par with capital and labour" (WEF, 2011: 7). The boom of big data technologies coincides with a period of prolonged economic recession. Ostensibly fulfilling Gillespie's statement, "the arrival of a new tool will often get entangled in current tales of social conflict and cultural failings, championed as a long-awaited solution" (2007: 2), industry stakeholders have assumed a position as proponents of big data as a remedy for the economy.

The World Economic Forum envisions data as an asset; "a person's data would be equivalent to their 'money.' It would reside in an account where it would be controlled, managed, exchanged and accounted for just like personal banking services operate today" (WEF, 2011: 10), supposedly giving individuals greater control and more adequate compensation for their data, and facilitating access to and the exchange of data between public and private enterprises (WEF, 2011: 10). The monetization of data aligns expectations with norms associated with other kinds of capital, invoking paradigms of property and commodity capitalism. The danger is that data, and by extension our identities, are being commodified, transforming identity into an economic asset, subject to market forces and privatization, rather than a manifestation of personal preferences and cultural traits.

In order to contextualize the European Commission's opposing view, it is essential to first examine the shifting technological and ontological context of big data.

When personal data becomes big data

Our contemporary lives are closely intertwined with technologies. Our senses are amplified by mobile technologies, our bodies are digitalized in electronic patient records and biometric identifiers and our intimate relationships are mediated by social networking sites. Our lives, leisure and work are interconnected by technologies in ever-expanding networks of humans and machines; "the world [has] become more connected, networked, and traceable" and as a result data sets of unprecedented complexity have become available (Rainie & Wellman, 2012 in Snijders et al. 2012: 1). Given the interweaving of technologies and daily life, the data that constitutes big data is undoubtedly personal and indeed the very nature of personal data, what can and cannot be protected by bodies such as the European Commission, is being warped by rising analytical capability (European Commission, 2012: 1). Indeed, technological innovation permits new possibilities of de- and re-identification of anonymized data, problematizing data security (WEF, 2013: 12). What is more as secondary usage of data becomes standard practice, and the re-combination of disparate data sets increases, representations of individuals multiply further.

From the data trail generated by mobile devices, it becomes possible to trace an individual's daily routine, habits, customs and preferences. Information about individuals' behaviour and location, once considered private, is known, and more significantly, commodified, by corporations receiving this data. The World Economic Forum echoes this assertion of analytical power; "mining and analysing this data give us the ability to understand and even predict where humans focus their attention and activity at the individual, group and global level" (2011: 5). The data generated through this co-operation with technology is gathered perpetually and often beyond individual control, resulting in a digital representation of an individual's life, distributed and disassembled in cyberspace. With big data capabilities, it becomes "possible to track social phenomena down to the individual level" (Lohr, 2013), and to use Pentland's words, with the ability to monitor and access insights from these "micro-patterns" we enter a "new era of social physics" (2012). As big data trends graduate towards this type of behavioural data, the implication of the lived-body in data generation is much greater. The potential for incorporating life itself into data analytics, and consequently into global markets, marks a noteworthy shift in how our lives, bodies and identities are implicated in big data.

The concept that our data can identify us or allow corporations to 'know' us and predict our actions calls into question what defines us as individuals and the potential impacts of big data on the future of identity. A special Eurobarometer report highlights tensions surrounding data disclosure: "74% of Europeans think that disclosing personal data is increasingly part of modern life, but at the same time, 72% of Internet users are worried that they give away too much personal data" (European Commission, 2012: 1). Whilst the statistics are relatively evenly matched, the nonchalant acceptance that continuous disclosure of data is "part of modern life" indicates a shift in attitude that could redefine our concept of identity, leading to the banalization of identity in society. As Davis and Patterson state, "if big data is evolving the meaning of the concept of identity itself, then big data is also evolving our ethical relationship to the concept the word represents" (2012: 16).

You are your data: rethinking the subject in the information era

In the era of big data analytics, not only are individuals recognized as both the subjects and producers of data, but also they *are* data. Dean states, 'people experience themselves as informationalized' (2002: 13), recalling the extent to which our identities co-evolve with technologies. Indeed, there is an "increasing mediation of our identities…through the interconnection of our bodies with information systems" (Van der Ploeg, 2003: 62). Data generated through this interconnection "constitute[s] digital representations of physical or bodily characteristics as individuals in one sense or another" and these representations "are functional in the construction and performance of our identities" (Van der Ploeg, 2003: 63). Because of the fundamental role of the embodied subject in data generation, be it active or passive, and that data's power to identify the individual it represents, it follows that the data cannot be conceived of as merely information.

As a result of digital technologies, parts of the embodied self circulate beyond the skin. Where in the past the physical body has signified a normative boundary of the self, the condition of virtuality, enabled by new technologies, dislocates the body from purely physical space and distributes it across time. Poster's work affirms this view; "in the mode of information, the subject is no longer located at a point in absolute time [and] space, enjoying a physical, fixed vantage point from which rationally to calculate its options. Instead it is multiplied by databases, dispersed by computer messaging and conferencing…dissolved and materialized continuously in the electronic transmission of symbols" (2001: 16). The information about a person's behaviour as a consumer, as a citizen and ultimately as a unique individual, constitutes a "digital persona" (Clarke, 1994 in Van der Ploeg, 2003: 70) which, despite its distributed nature, is "constitutive of and inseparable from" embodied identity (Van der Ploeg, 2003: 70). The informational body challenges conventional notions of subjectivity and embodiment, and it is this body ontology that underpins the context of big data.

The European Commission's vision of personal data serves as a springboard for conceptualising a different approach to big data: an embodied approach. The European Commission portrays personal data as part of the self, encouraging individuals to assert more care over their online activity stating, "your data, it's you" (2013). The video, which promotes changes to the Data Protection Directive, entitled "Take control of your personal data" (EUJustice, 2012), brings the body to the forefront of attention. The 50-second video has no speech and deploys quick cuts between three scenes depicting individuals online at work, at home and in a public space. Then, in the following shots of each individual, their clothes are whisked from their bodies leaving them entirely naked. The advert concludes with the lines: "Online you reveal more than you think. Take control of your personal data" (EUJustice, 2012). By translating the online world of avatars, handles and relative anonymity into the offline context of being naked in public, the European Commission makes an implicit suggestion about the equal vulnerability of our online and offline self. The comparison with a normative notion of bodily integrity calls to the fore the complex circumstances of managing personal data in virtual contexts.

Other theorists have also leveraged the concept of nudity in order to attenuate the distinction between online and offline norms of privacy and identity. For example, Bartow presents a second interpretation of the marketing trend "shop naked" (2000: 648), originally meant to encourage consumers to shop online, thereby unencumbered by bags, queues, children, parking etc. However Bartow's reading is more sinister, envisioning online consumers as "figuratively nude", stating, "Netscape and its clients know everything about us, our personal flaws and most intimate attributes have been stripped from us and are known to all" (2000: 648). Bartow emphasizes the vulnerability of our intimate information online through the concept of nudity, re-placing the physical body where it is often forgotten in virtual interactions. By drawing on bodily vulnerability, individuals are urged to reconsider their online values and behaviour from an embodied perspective, more often associated with the offline world.

The fast accelerating direct and indirect generation of data that identifies us as individuals, and the lack of ability to control it, is a problem that requires further debate because it stands in tension with norms of integrity and privacy. The collective concern for how we represent ourselves to the world is the essence of the moral right to privacy and personal integrity. Alterman defends the right to privacy as the right to exercise "control over how and when we are represented to others" (2003: 143). In the context of big data however, representations of the self multiply; through participation in online networks, mobile device use and credit card purchasing, individuals create, actively and passively, a detailed and targetable representation of their self. This is further compounded by the possibilities of secondary use of data, particularly in combination with otherwise separate data sets, such as associating medical records with insurance data. Alterman argues, "the proliferation of representations that identify us uniquely thus involves a loss of privacy, and a threat to the self-respect which privacy rights preserve" (2003: 143), highlighting the importance of representations of the self, both virtual and physical, to the notion of identity. Personal data constitutes an essential representation of the self, and when combined with the power of big data analytics, these representations can target users on an individual level. Thus, the European Commission's envisioning of personal data as part of 'you' creates an instructive starting point for reconceptualising data in contrast with industry discourse that champions data as an economic asset. As representations of the self proliferate, individuals must reconceive the delineation and demarcation of the self in order to assert their rights to identity and integrity.

Embodiment Theory

This paper proposes an embodied approach in which the data created by and about Internet users constitutes a fundamental part of the self. In order to interrogate the philosophical underpinnings of the relationship between big data and the body, Merleau-Ponty's phenomenology (1962) will be used to inform a model of embodiment that encompasses digital technologies as part of the embodied self. By reiterating embodiment theory in the contemporary context of big data, we can challenge the assumptions about what constitutes the embodied subject upon which industry discourse is based and can prompt a more holistic perspective on big data.

Phenomenological philosophy is predicated on a malleable and extendable body, capable of seamlessly encompassing tools and technologies into the ensemble that constitutes the self. In phenomenology, the body is not reduced to its material essence but reconceived as the incorporation of mind, body and technological extensions within a corporeal schema that is in a perpetual process of becoming-in-the-world. Merleau-Ponty's recognition of a corporeal schema, open to extensions by means of technology, originated well before the condition of virtuality and digital presence would enter the scene to further problematize bodily boundaries. Yet his advocacy for the plasticity of the body makes reinstating his philosophy in the context of big data particularly apt.

Merleau-Ponty employs the term "corporeal schema" to encompass the transforming and shifting boundaries of the physical body extended and morphed by technological tools, "rendering external objects as internal and projecting a body-image that is in continuous flux to incorporate new instruments" (Ajana, 2005). The open and interconnected nature of our being-in-the-world proposed by phenomenological thought correlates with the way in which contemporary lives are increasingly interconnected with external tools and technologies. Merleau-Ponty describes the way in which the subject integrates external instruments into the corporeal schema:

> The blind man's stick has ceased to be an object for him, and is no longer perceived for itself; its point has become an area of sensitivity, extending the scope and active radius of touch, and providing a parallel to sight. In the exploration of things, the length of the stick does not enter expressly as a middle term… To get used to […] a stick is to be transplanted into [it], or conversely, to incorporate [it] into the bulk

> of our own body. Habit expresses our power of dilating our being-in-the-world, or changing our existence by appropriating fresh instruments. (1962: 143)

The 'dilation' of one's being-in-the-world is testimony to the malleability of bodily boundaries, the ability to incorporate external objects and thereby reconfigure our being-in-the world contextually. Merleau-Ponty's concept of the corporeal schema, open to technological extensions, lays the foundations for the extension of the body through digital representations. Indeed, in the context of media theory, McLuhan deemed media technologies the "extensions of man"; he states "to behold, use or perceive any extension of ourselves in technological form is necessarily to embrace it… to accept these extensions of ourselves into our personal system" (1964: 68). Echoing Merleau-Ponty[1], both authors exalt the body's capacity to seamlessly incorporate technology into corporeal activities. In an embodied approach to big data this corporeal schema becomes the subject of consideration, addressing the individual as both a human and data subject.

Richardson and Harper (2001) develop a useful perspective on Merleau-Ponty's philosophy in the context of digital technologies, which I propose to take forward in the context of big data. First, the authors assert the "fundamental irreducibility between technologies and embodiment' upheld by phenomenology in which tools and technologies are 'not conceived as merely perceptual attachments or extensions, but rather are incorporated into our embodied field or corporeal schema" (2001: 9). This 'fundamental irreducibility' is essential to our consideration of the relationship between digital representations of the self and issues of identity and selfhood, which are often bifurcated in conventional discourses of virtuality. Whilst the discussion of tools and technologies seems to allude to physical extensions of the corporeal schema, it is essential to recall that with the rise of digitalization and informatization, these 'tools' can also be virtual. Richardson and Harper affirm, "the phenomenological and corporealist approaches thus offer a particularly useful foundation for developing a model of embodiment relations which can account for the way in which the virtual is both a technological ensemble and an always-already a corporeal condition" (2001: 9).

1 Though it is not evident whether the two theorists were aware of one another. See S. Basbaum (2011) for a more detailed investigation of McLuhan and Merleau-Ponty.

The concept of the two-fold reality of the virtual can be translated into a reconceptualization of ontology in big data. Virtuality is not wholly disembodying but rather another technology with which we choose to extend the body, thus necessitating "a shift from thinking of the virtual as de-corporealizing subjectivity, toward an understanding of how the body incorporates the virtual" (Innis 1984, in Richardson and Harper, 2001: 10). Richardson and Harper's conceptualization of virtuality is key to thinking about data technologies as "both a technological ensemble and an always-already a corporeal condition" (2001: 9). A reconceptualization of big data through a phenomenological lens emphasizes the interconnected and fundamental relationship between the body and technology. Bringing to light the embodied nature of big data compels the recognition of a deeper complexity in the relationship between data, the body and identity, thereby demonstrating that the definition of big data as an economic asset is not so straightforward. The final section will discuss how and why this conception of data is problematic by investigating the discursive context of commodity capitalism and the processes of fragmentation that underpin this ideology.

Understanding (dis)embodiment in digital worlds

Discourses about the body in virtuality influence the way we think about bodies in big data. Narratives of absolute disembodiment in both popular and critical registers bolster the collective imaginary of virtual reality and the role of the body therein. For example, Barlow's pivotal publication, "The Declaration of Independence of Cyberspace", portrays cyberspace as "the new home of Mind... a world that is both everywhere and nowhere, but it is not where bodies live" (2001: 28). The capitalization of 'Mind' and the dropped article denotes an air of supremacy, evocative of Cartesian transcendental objectivism. Similarly, for cyberspace philosopher Michael Heim, in virtual worlds "minds are connected to minds, existing in perfect concord without the limitations or necessities of the physical body" (1993 in Richardson and Harper, 2001: 1). The way in which the mind seemingly detaches from the body resonates with this neo-Cartesian vision of the body in big data: the conceptualization of data as an asset relies of the alienation of data from the embodied subject, thereby freeing it to circulate as a commodity. Furthermore, narratives of disembodiment propel a reductionist epistemology that stems from out-dated concepts of mind/body dualism, concurrently encouraging the abandonment of the body and all of it's social, cultural and ideological realities. It is against this discursive backdrop, dominated by discourses that negate the body, that digital identities proliferate in data.

By contrast, phenomenological philosophy advocates "the body is our general medium for having a world" (Merleau-Ponty, 2007: 13), thus asserting the centrality of corporeal ontology no matter the context. This focus on corporeality is essential not only to rethinking the limits of the corporeal schema in the era of big data but also it can instigate a more ethical perspective. Frohmann's work on cyber-ethics draws on MacIntyre's theory of "acknowledged dependence", asserting that the foundation of our sense of ethics stems from bodily vulnerability (2000). This human fragility binds individuals to one another in a relationship of interdependence. Noting the prevalence of cyber-discourses that negate the body in a contemporary resuscitation of Cartesian dualism, Frohmann proclaims that information ethics requires an approach that envisions "information as an effect of the relevant social practices of embodied subjects" (2000: 432). He states, "to address serious moral issues, information ethics cannot abstract from our bodily, animal nature" (Frohmann, 2000: 423). To disregard the embodied subject is to disregard the body as both an anthropological signifier and as an ethical agent. An embodied approach upholds an anthropological and ethical gaze, which can enlighten and enliven the conceptualization of big data, particularly in relation to industry's understanding of data as an asset.

Conceptualising data as an asset

It is essential to interrogate what assumptions are being made, particularly at the top of industry and at a pan-national level, regarding what constitutes big data because how we define big data and sanction activities involving it bears consequence for the common notions of identity and embodiment. An embodied approach demonstrates how and why the propertization of big data is problematic: not only that the individual cannot be separated from data so unproblematically but also that these processes of fragmentation are unethical and dehumanising. The conceptualization of data as an economic asset overlooks the significance of big data as both a "technological ensemble" and a "corporeal condition" (Richardson and Harper, 2001: 9), instead alienating data from the subject by aligning conceptions of data to a framework of capitalistic values.

Conceiving data as an economic asset incurs a process of fragmentation of the embodied self and alienation from the individual it represents in order for it to circulate unproblematically as a commodity in the circuit of capital. The discourse deployed by the World Economic Forum situates big data and personal data in the context of capital; for example, "individuals are being understood as both producers and consumers of data" (2013: 18). Conceptualizing personal data as an asset, "a new type of raw material that's on par with capital and labour" (WEF, 2011: 7), instigates a discursive split between the individual and the intimate information that comprises their data, which is "constitutive and inseparable from" the embodied self (Van der Ploeg, 2003: 63).

The fragmentation of the embodied self and the disregard for the corporeal ontology of big data is significant because "the commodity fetishism of capitalism… derives from the alienation that arises between persons and the things that they produce that are then put into circulation in the monetary economy" (Lock, 2007: 571). In the conceptualization of data as an asset, intimate personal information seems to unproblematically separate from the individual and thereby circulate as part of the economy; this alienation underpins the "phenomenology of the commodity as a self-enclosed entity, dominant over its creators, autonomous, and alive with its own power" (Taussig, 1980 in Lock, 2007: 571). The evocation of the autonomy and dominance of commodity, when considered in the context of big data, echoes concerns about the agency that data acquires to circulate and communicate despite its embodied origin (Aas, 2006). It can be argued that industry bodies such as the World Economic Forum purposefully play on the alienation between individuals and the commodities they produce, normalized in the context of contemporary capitalism, in order to encourage increased data circulation and sharing.

Indeed, this hypothesis is clearly reflected in an examination the World Economic Forum's priority, trust. Key points outlined in the report regarding appropriate consent mechanisms, data protection and security are explicitly designed to "build trust into the system, and encourage a reliable, predictable and more valuable flow of data into and within the system" (2013: 17). This concern for trust is impelled by the underlying objective of generating greater value from data generated by individuals. The World Economic Forum seeks to build a trustworthy environment that will encourage the free sharing of data, in other words, the free flow of capital. Outlining their principles for the "global trusted flow of data" (2013: 11), the World Economic Forum discusses measures of "accountability and enforceability needed to cultivate trust… [thereby] persuading individuals to be more willing to share data about themselves" (2013: 15). The explicit expression of objectives that solely comprise the increased circulation of data assets, without any concern for the individual implicated, is a cause for concern. The World Economic Forum fails to explicitly confront fundamental implication identity in big data, instead regarding the individual as no more than a gatekeeper on a desired resource.

Let us draw a parallel with the context of the commercialization of DNA practices, a field in which the embodied subject is also informationalized and digitized. Lock states, "in order for body parts to be made freely available for exchange they must first be conceptualized as thing-like, as detachable from the body" (2007: 572). Intellectual property law on genetic patenting reveals a normative dichotomy of the biological and the technological, which legally distinguishes what can and cannot become property: "in order to procure a patent on a biological it must be shown that through the 'process of their production' the 'natural' object has been transformed into an 'invention' " (Lock, 2007: 574). What can legally become an asset is thus defined based on normative notions that separate the body and technology. By retreating into a vision of data as acorporeal, immaterial and abstracted from the body, it is more easily envisioned as a tradable asset. Yet there is a "fundamental irreducibility" (Richardson and Harper, 2001: 9) between humans and data, in both DNA and big data, therefore identifying this discontinuity is counterintuitive. Van der Ploeg's argument against the dichotomy of the body itself and bodily information asserts that the body is always and already information; "there is no clear point where bodily matter first becomes information" (2003: 70). This assertion resonates with the embodied approach, which sustains a vision of data as both a "technological ensemble" and an "always-already corporeal condition" (Richardson and Harper, 2001: 9), thereby bringing digital representations of the subject into consideration as part of the embodied self.

Only when data is disembodied can corporations lay claim to it and then the individual more willingly exchanges and trades his or her personal data, forgetting or perhaps never realising its intimate and bodily origins.

Whilst it could be argued that individuals are all homogenous when reduced to bits and bytes of data, to conceptualize data as an asset is to dehumanize it, disregarding the individuals, identities and bodies that constitute the subject of big data. By consequence of the bifurcation of the biological and the technological, and the subsequent reification of data, the individual more willingly exchanges and trades his or her personal data, forgetting its intimate and bodily origins. This process of dehumanization facilitates the entrance of data into circuits of capital as a commodity. By contrast, deploying an embodied approach to consider data as part of the self asserts the importance of a corporeal ontology and urges a holistic approach in which data subjects are contextualized and ultimately, embodied.

The commodification of life itself

The commercialization of big data as an asset, bearing in mind the implication of bodies and identities inherent in data, blurs the concept of what, or *who*, is the commodity in the circuit of capital. The World Economic Forum's conceptualization of big data as an economic asset, viewed through an embodied perspective, can be interpreted not only as the commodification of identity, but also as the commodification of life itself. Conceiving data as an asset is predicated on the assumption that data is acorporeal, obscuring the reality of individuals as co-creators of data in question. Personal data now makes up the vast majority of big data (WEF, 2011: 9); the analysis and monetization of our habits, tastes and locations is, in other words, the commodification of life itself.

Kember and Zylinska remark this phenomenon in the context of social networking sites, a key platform for major voluntary data sharing and passive data collection: "as carriers and signallers of valuable information, such platforms turn not just us as their users but also 'life itself' into a product, constantly remoulded and repacked via the flickering pulsations and pings of data" (2012: 163). To similar effect, Dean's theorization of "communicative capitalism" (2009, 2002) argues against the networks of technologies that extract value from social relationships and communication (2002: 4). She notes the socio-cultural anxieties of commodification: 'the commodification of communication reformats ever more domains of life in terms of the market: What can be bought and sold? How can a practice, experience, or feeling be monetized?" (2009: 25), depicting a foreboding image of the future for emotional and social experience.

Conclusion

In conclusion, the juxtaposition of these opposing perspectives brings to light the complexity of the relationship between data, the body, identity and commodity capitalism and creates space for critical reflection on the philosophical and ideological underpinnings of the relationship between big data and the body in their social totality. Conceptualising data from an embodied perspective enlivens the body and empowers the individual, as user, producer, in a relationship of co-constitution with data. The discussion of the body as property, in terms of alienation and commodity capitalism, brings to light the boundaries of how we define and delineate human identity and its value. Whilst phenomenology is instructive in bringing to light the embodied nature of big data, this reconceptualization of phenomenology also brings to light its limitations in the context of the data revolutions.

Unprecedented uses of data in which bodies and identities are implicated are permitted by today's big data analytics and therefore our very notions of embodiment and identity require critical reflection. As representations of the self multiply through digital platforms and data collection, the phenomenal body is extended into virtual worlds and the boundaries of the corporeal schema dilate on an unprecedented scale, far beyond the current precedent of what is conventionally accepted as being part of the self. It could be argued that phenomenology is not a sufficiently encompassing maxim in this new context because the data that exists about individuals circulates autonomously and beyond our control, indicating a fundamental ontological shift in the make up of our identities. Whilst the body remains central to existence and experience, as phenomenological philosophy asserts, in the context of big data it becomes necessary to question the limits of phenomenology, as we know it. How to reconceive the self when the information subject is multiplied and dispersed through cyberspace is a central consideration. Indeed, the evolving perception of bodily boundaries poses complex challenges to long-standing metaphysical assumptions about what it means to be a body, or indeed a human in the world.

References

Aas, Katja Franko, 2006, "The body does not lie: Identity, risk and trust in technoculture", in *Crime Media Culture* 2(2), pp. 143–158.

Ajana, Btihaj, 2005, "Disembodiment and Cyberspace", in *Electronic Journal of Sociology,* Retrieved from <http://www.sociology.org/content/2005/tier1/ajana.html>.

Alterman, Anton, 2003, "A Piece of yourself: Ethical issues in biometric identification", in *Ethics and Information Technology,* 5(3), pp. 139–150.

Barlow, John Perry, 2001, "A Declaration of the Independence of Cyberspace", in Ludlow, Peter (ed.), *Crypto Anarchy, Cyberstates, and Pirate Utopias,* Cambridge, MA: MIT Press, pp. 27–30.

Bartow, Ann, 2000, "Our Data, Ourselves: Privacy, Propertization, and Gender", in *University of San Francisco Law Review,* 34 (Summer), pp. 633–704. Retrieved from <http://ssrn.com/abstract=374101>.

Basbaum, Sergio, 2011, "McLuhan and Continental Philosophy: How Merleau-Ponty's Phenomenology of Perception helps support and continue McLuhan's directions", in Ciastellardi Matteo et al. (eds), *Conference Proceedings of Understanding Media, Today: McLuhan Galaxy Conference. Barcelona, May 2011,* SP: Collection Shehen, Editorial Universidad Oberta de Catalunya, pp. 552–561.

Bassett, Caroline, 2007, *The Arc and the Machine: Narrative and New Media,* Manchester, Manchester University Press.

Conatser, Trey, 2010, "There's no 'I' in information: some naysayings for new media studies", in *New Media & Society,* 12(3), pp. 365–378.

Davis, Kord and Patterson, Doug, 2012, *Ethics of Big Data*, Sebastopol, CA: O'Reilly Media.

Dean, Jodi, 2009, *Democracy and Other Neoliberal Fantasies*, Durham and London, Duke University Press.

Dean, Jodi, 2002, *Publicity's Secret: How Technoculture Capitalizes on Democracy*, Ithaca & London: Cornell University Press.

European Commission (2013) *Protecting your data, protecting you*, [Online] European Commission. Retrieved from <http://ec.europa.eu/justice/data-protection/index_en.htm>.

European Commission (2012) *Why do we need an EU data protection reform?,* [Online] European Commission. Retrieved from <http://ec.europa.eu/justice/data-protection/document/review2012/fact-sheets/1_en.pdf>.

EUJustice, 2012, *Take control of your personal data*. Retrieved from <https://www.youtube.com/watch?feature=player_embedded&v=5ByVaZ0rg8U>.

Frohman, Bernd, 2000, "Cyber-Ethics: Bodies or Bytes?", in *International Information & Library Review,* 32, pp. 423–435.

Gillespie, Tarleton, 2007, *Wired Shut: Copyright and the Shape of Digital Culture*, Cambridge, MA: MIT Press.

Kember, Sarah and Zylinska, Joanna, 2012, *Life After New Media: Mediation as a Vital Process*, Cambridge, MA and London: MIT Press.

Lock, Margaret, 2007, "Alienation of Body Parts and the Biopolitics of Immortalised Cell Lines", in Lock, Margaret and Farquhar Judith (eds), *Beyond the Body Proper*, Durham and London, Duke University Press, pp. 567–583.

Lohr, Steve, 2013, "The Promise and the Peril of the Data-Driven Society", in *The New York Times*. Retrieved from <http://bits.blogs.nytimes.com/2013/02/25/the-promise-and-peril-of-the-data-driven-society/?_r=0.2013>.

McLuhan, Marshall, 1964, *Understanding Media*, Berkeley, CA: Gingko Press Inc.

Merleau-Ponty, Maurice, 1962, *The Phenomenology of Perception*, London, Routledge.

Pentland, Alex, 2012, "Reinventing Society in the Wake of Big Data", in *Edge*. Retrieved from <http://www.edge.org/conversation/reinventing-society-in-the-wake-of-big-data>.

Van der Ploeg, Irma, 2003, "Biometrics and the body as information: Normative issues of the socio-technical coding of the body", in Lyon, David (ed.), *Surveillance as Social Sorting*, London, Routledge, pp. 57–73.

Poster, Mark, 2001, *The Information Subject*, London, Routledge.

Richardson, Ingrid and Harper, Carly, 2001, "Corporeal Virtuality: The Impossibility of a Fleshless Ontology", in *Body, Space and Technology Journal*, 2(2). Retrieved from <http://people.brunel.ac.uk/bst/vol0202/index.html>.

Snijders, C., Matzat, U. and Reips, U. (2012). "Big Data: Big gaps of knowledge in the field of Internet", *International Journal of Internet Science*, 7, 1–5.

Tennenhouse, David, 2013, "The Data Dilemma", in *Microsoft on the Issues: News and perspectives on legal, public policy and citizenship topics*. Retrieved from <http://blogs.technet.com/b/microsoft_on_the_issues/archive/2013/02/28/the-data-dilemma.aspx>.

World Economic Forum, 2013, *Unlocking the Value of Personal Data: From Collection to Usage*. Retrieved from <http://www3.weforum.org/docs/WEF_IT_UnlockingValuePersonalData_CollectionUsage_Report_2013.pdf>.

World Economic Forum (2011). *Personal Data: The Emergence of a New Asset Class*. Retrieved from <http://www3.weforum.org/docs/WEF_ITTC_PersonalDataNewAsset_Report_2011.pdf>.

List of Contributors

Aysun Akan
Faculty of Communication
Izmir University of Economics
Turkey

Richard Balme
Paris School of International Affairs
Sciences Po, Paris
France

Josie Bernicot
CeRCA-CNRS, UMR 7295, MSHS
Université de Poitiers
France

Alain Bert-Erboul
CeRCA-CNRS, UMR 7295, MSHS
Université de Poitiers
France

Zhe Deng
School of Public Policy and Management
Tsinghua University, Beijing
China

Alain-Philippe Durand
Department of French and Italian
University of Arizona
United States of America

Antonine Goumi
CHArt UPO EA 4004
Université Paris Ouest Nanterre La Défense
France

Marina Haan
GRIC (Groupe de Recherches Identités et Cultures)
Université du Havre - IUT
France

Abhay vir Singh Kanwar
Department of Humanities and Social Sciences
Indian Institute of Technology of Bombay
India

Fabien Liénard
IDEES Le Havre, UMR CNRS 6266
Université du Havre - IUT
France

Danielle Masterton
King's College London
United Kingdom

Sara Minucci
OMERA Social Sciences
Torino World Affairs Institute
Italy

Elyse Petit
Department of French and Italian
University of Arizona
United States of America

Pitabas Pradhan
Department of Mass Communication
Aligarh Muslim University
India

Marta Severo
Laboratoire Geriico
Université de Lille 3
France

Kirsi Viitanen
CICERO Learning Network
University of Helsinki
Finland

Marianna Vivitsou
CICERO Learning Network
University of Helsinki
Finland

Olga Volckaert-Legrier
Octogone-ECCD EA 4156
Université Toulouse Jean Jaurès
France

Yves Winkin
Ecole Normale Supérieure of Lyon
Conservatoire National des Arts et Métiers of Paris
France

Sami Zlitni
IDEES Le Havre, UMR CNRS 6266
Université du Havre - IUT
France